DEBATING SCIENCE

DEBATING SCIENCE

DELIBERATION, VALUES, AND THE COMMON GOOD

EDITED BY

DANE SCOTT AND BLAKE FRANCIS

Humanity Books

an imprint of Prometheus Books
59 John Glenn Drive, Amherst, New York 14228-2119

Published 2012 by Humanity Books, an imprint of Prometheus Books

Cover design by Liz Scinta

Inquiries should be addressed to
Humanity Books
59 John Glenn Drive
Amherst, New York 14228–2119
VOICE: 716–691–0133
FAX: 716–691–0137

16 15 14 13 12 5 4 3 2 1

Library of Congress Cataloging-in-Publication Data

Debating science : deliberation, values, and the common good / edited by Dane Scott and Blake Francis

p. cm.

Includes bibliographical references.

ISBN 978–1–61614–499–9 (pbk. : alk. paper)

ISBN 978–1–61614–500–2 (ebook)

1. Science—Moral and ethical aspects. 2. Technology—Moral and ethical aspects. I. Scott, Dane, 1957– II. Francis, Blake, 1978–

BJ57.D43 2011

170—dc23

2011040626

Printed in the United States of America on acid-free paper

CONTENTS

Preface 9

PART ONE: FOUNDATIONAL ISSUES

Section 1. Ethics and the Science Debates

1. Debating Science: Ethics Education and Deliberation
   Dane Scott 17

2. Overcoming Scientific Ideology
   Bernard Rollin 33

3. Overcoming Philosophobia: A Few Ethical Tools for the Science Debates
   Christopher J. Preston 49

4. Bridging the Gap: Global Justice in Health Research
Julian Culp and Nicole Hassoun 69

Section 2: Policy and the Science Debates

5. Intellectual Liberty and the Public Regulation of Scientific Research
Clark Wolf 83

6. Efficiency vs. Equity: Economic Considerations in the Science Debates
Richard Barrett 101

7. Communicating Science: Moral Responsibility in Theory and Practice
Wendy S. Parker 117

PART TWO: SCIENCE AND TECHNOLOGY DEBATES

Section 1. Biotechnology

8. Is Sustainability Worth Debating?
Paul B. Thompson 133

9. Biotechnology and the Pursuit of Food Security
David Castle 147

Section 2. Climate Change

10. Ethically Dealing with Climate Change: Comparing the Maldives, China, and the United States
Bill McKibben 163

11. Science, Ethics, and Technology and the Challenge of Global Warming
Albert Borgmann 169

12. Ten Ethical Questions That Should Be Asked of Those Who Oppose Climate Change Polices on Scientific Grounds
Donald A. Brown 179

Section 3: Nanotechnology

13. Value-Sensitive Design and Nanotechnology
Ronald Sandler 205

14. Debating Nanoethics: US Public Perceptions of Nanotech Applications for Energy and Environment
Barbara Herr Harthorn, Jennifer Rogers, Christine Shearer, and Tyronne Martin 227

Contributors 245

Notes 251

PREFACE

Because of the many benefits of modern science and technology, people in industrialized countries have become habituated to looking to science and technology to solve problems. However, scientific and technological "progress" is no longer taken as an unqualified good. "Technological fixes" are viewed with deep suspicion. This leads to a certain incongruity in modern thought; people increasingly look to science to solve problems even while many are increasingly critical of using science to solve problems. This, of course, can be easily explained. Science and technology have made us healthier, allowed us to grow more food, and created fascinating and useful inventions that nearly exhaust our imaginations. However, at the same time, science and technology can have severe health and environmental side effects and unintended consequences, leading many to question whether our modern technological lifestyle is really the good life. Because of this, some greet new developments like transgenic crops and genetically modified organisms with hostility. In addition, scientific research, like climate science, has come under intense criticism largely for political reasons. Nonetheless, over the next fifty years the rising generation of global citizens must successfully navigate the climate crisis and feed an additional 3 billion people sustainably; science and technology will, and must, play a role in solving these problems. We can expect the role of science and technology to increasingly be the subject of intense public debate.

These debates will have far-reaching ethical implications. Unfortunately, by most accounts, over the last thirty years or so science debates have not gone well. Chances are poor-quality debates will not lead to wise decisions or happy outcomes. For democratic societies to fully and appropriately utilize science and technology to solve problems of a global magnitude, we must improve the quality of public discourse. The stakes are too high for conflict and strategic discourse to dominate decisions of far-reaching consequences for the planet and civilizations. The contentious nature and low quality of science debate indicates a breakdown in ethics and public discourse. This book seeks to create an opportunity for readers to seriously consider what it might take for us to become more thoughtful deliberators and engaged global citizens.

The goals of this anthology, then, are to provide readers with the opportunity to become better informed about the practice of cooperative deliberation and to critically examine the ethical and moral aspects of science, technology, and public discourse. To fulfill these goals, a variety of salient issues are covered. Part 1 of this anthology is divided into two sections that cover foundational issues. The first section, "Ethics and the Science Debates," consists of four essays that address overcoming obstacles to the inclusion of ethics and political philosophy in thinking about the directions of scientific research. The second section of part 1, "Policy and the Science Debates," contains discussions about four important topics: the regulation of scientific research and emerging technologies, the scope and limits of scientific freedom, the role of economics in science debates, and the moral responsibilities of scientists to communicate scientific findings to the public. Part 2 is divided into three sections. Each section covers an important recent science and technology debate: agricultural biotechnology, climate change, and nanotechnology. While the authors discuss specific science and technology debates, they draw general conclusions about incorporating ethical and moral considerations in public deliberations over science and technology. The chapters in the agricultural biotechnology section provide an ethical context for debates over agricultural biotechnology by discussing the normative goals of sustainability and food security. The next section contains three essays on the debate over global climate change. The first two essays are insightful evaluations of the political responses to the science of climate change. The third essay, Donald A. Brown's "Ten Ethical Questions That Should Be Asked of Those Who Oppose Climate Change on Scientific Grounds," covers ethical issues arising from scientific uncertainty in climate science. The final section is devoted to two essays on nanotechnology. In the first, "Value-Sensitive Design and Nanotechnology," Ronald Sandler offers suggestions about incorporating values in design decisions. The final essay, "Debating Nanoethics: US Public Perceptions of Nanotech Applications for Energy and Environment" by Barbra Herr Harthorn, Jennifer Rogers, Christine Shearer, and Tyronne Martin, wraps up many themes in this volume by investigating public perceptions of key issues in debates over emerging technologies and focusing on shared values associated with the research, development, and commercialization of nanotechnology.

This anthology is the product of the Debating Science Program, supported by the National Science Foundation (NSF), Ethics Education for Graduate Students in Science and Engineering (EESE) program. This work was supported by, NSF, SES-062950. We thank Christopher Preston and Rebecca Bendick, who (along with Dane Scott) served as co-PI's for the Debating Science Program and assisted in developing the 2007 and 2008 workshops and online teaching components. In addition, we thank Christopher Preston for his additional contributions to this volume. We thank the Mansfield Ethics and Public Affairs Program at the University of Montana, UM president Royce Engstrom, and provost Perry Brown for supporting the Debating Science Program. We are grateful to the center's staff, Kari

Samuel and Justin Whittaker, for their excellent work in organizing and implementing the Debating Science workshops and online environments. We acknowledge the many outstanding scholars who contributed to the Debating Science workshops. Most importantly, we appreciate the contributions of the talented graduate students who participated in the program. We thank Özlem Özgür for donating her time to create the tables, graphics, and illustrations for this book. Dane Scott would like to thank his family, Leslie, Sam, and Katy. Finally, Blake Francis would like to thank his parents, Jim and Mary, for their support and encouragement over the years.

Part One

FOUNDATIONAL ISSUES

Section 1.

ETHICS AND THE SCIENCE DEBATES

1.

DEBATING SCIENCE

Ethics Education and Deliberation

Dane Scott

SCIENCE AND THE CHAOTIC ARENA OF PUBLIC DEBATE

Science and technology are increasingly thrown into the chaotic arena of public debate. Since the public has much at stake, this is probably as it should be. Unfortunately, by most accounts these debates have gone poorly. For example, the many issues surrounding global climate change have been the subjects of polarizing disagreement for decades. Innovations in agricultural biotechnology have been hotly disputed for over twenty years. And powerful new developments in nanotechnology are often being met with similar treatment: contentious, adversarial debate. For democratic societies to fully and appropriately utilize science and technology to solve global problems, it would be wise to improve the quality of public discourse. The stakes are too high for conflict and manipulative discourse to dominate these important decisions.

The poor quality of public debates indicates severe gaps between science, ethics, and public discourse. To help fill these gaps an explicit focus is needed to assist students and the public to become more thoughtful deliberators and engaged global citizens. In what follows, I will briefly introduce four areas of study and research that will help fill this need: (1) ethics education, debate and deliberation, (2) systematic and coherent deliberations, (3) group dynamics and social influences on deliberations, and (4) deliberation, practical reason, and ethical theory.

ETHICS EDUCATION, DEBATE, AND DELIBERATION

This chapter argues that the emphasis on ethical deliberation provides an important countermeasure to the ethos of adversarial debate in popular and political cultures. The term, "ethical deliberation," is a way of emphasizing a systematic effort to include a wide range of value considerations in the deliberative process, as opposed to merely practical or instrumental concerns. Popular media actively encourage a culture of polarizing debate that is often lamented even while being promoted. For example, debate dominates political television programming, where up to six talking heads will pop on a split screen and go at each other on cue. This spectacle might easily be followed by an editorial where a pundit bemoans our polarized and ineffectual political processes. Rather than reinforce these negative trends in public discourse, this chapter and this anthology seek to provide readers with food for thought for considering an alternative practice of ethical deliberation. This consideration takes place in the context of some of the most important and controversial scientific research and emerging technologies of our time, for example, medical research, animal welfare, scientific liberty, global climate change, agricultural biotechnology, and nanotechnology.

Of course, adversarial debate has a place in public and academic discourse, and it can be productive, engaging, and entertaining. But its dominance implicitly teaches that adversarial debate is *the* type of dialogue for resolving controversial issues. This leads to the unfortunate consequence that as a society we often debate when we should deliberate.

The philosopher Douglas Walton's work in informal logic, or logical pragmatics, shows that debate and deliberation are distinct, if sometimes intermixed, dialogical activities.[1] In a debate, opposing sides attempt to persuade a third party that they are right and their opponents are wrong, often by whatever means works. Deliberation involves a group of concerned citizens with common interests and shared fates. This group takes on the task of defining problems, characterizing goals, assessing alternative means to reach those goals, and, hopefully, agreeing upon a course of action. On one hand, debate begins with opponents advancing competing positions with dogmatic certainty. On the other, deliberation begins with the question: What should *we* do to solve *our* problems and reach *our* goals? Public discourses in controversial areas like nanotechnology, climate change, and agricultural biotechnology all aim at deciding on courses of action to solve problems and reach goals. Deliberative dialogue is more appropriate than debate for collective decisions in which no one can be certain of the way forward and everyone has a stake in the outcome.

PROBLEMS WITH THE CULTURE OF ADVERSARIAL DEBATE

The idea that we are debating when we should be deliberating is illustrated by the controversy surrounding global climate change. In their book *The Science and Politics of Global Climate Change*, Andrew Dessler and Edward Parson write:

> The climate change debate, like all policy debates, is ultimately an argument over action. How shall we respond to the risks posed by climate change? . . . Listen to the debate over climate change and you will hear people making many different kinds of arguments—about whether and how the climate is changing, whether human activities are affecting the climate, how the climate might change in the future. . . . Their goal is to convince others to support a particular course of action.[2]

The climate change debate in the United States has been deadlocked for well over two decades. This near stalemate is placing the world in ever-increasing risk of catastrophic climate change and increasing the difficulty of ever effectively addressing this problem. One contributing reason for this stalemate may be the ethos of adversarial debate that dominates public discourse. Solutions to the climate crisis, for instance, will require cooperative intelligence and collective action. While it would be impossible to make sweeping changes in the larger culture, it is important for educational programs in research and applied ethics in particular, and higher education in general, to encourage a deliberative culture. In the rest of this section, I will provide several reasons deliberation, *not* debate, is the most appropriate from of dialogue about these kinds of issues involving science and technology.

The debate format can misrepresent reality on many issues where science and technology play a key role. One reason for this is that competing sides enter the debate confident that they know the way forward. For example, climate change is a dynamic problem of unprecedented scale with many uncertainties. No one person or group can know what should be done with the kind of dogmatic certainty that is commonly found in adversarial public debates. Climate change and the future of agriculture, for instance, are complex and multidimensional issues, involving science, technology, ethics, politics, economics, and public policy to name a few areas. The crude framing of such complex issues in terms of pro and con, for and against, is a serious mistake that can lead to dangerous outcomes. Yet it is common for people to line up and label themselves as believes or unbelievers in climate change, or as pro-GMOs (genetically modified organisms) or anti-GMOs. This is done as if assessments of scientific information and decisions about new technologies were akin to religious or political affiliations. Nonetheless, simplistic black and white thinking on issues of great consequence and uncertainty begs sweeping answers that dangerously miss the subtleties of reality.

Issues like climate change, the future of agriculture, and global health all involve countless uncertainties, unknown contingencies, and profound risks that will play out well into the future. Well-designed and well-intentioned plans often go wrong. No one can dogmatically argue about what to do about these kinds of issues. Yet the debate format implies certitude, in which both sides boldly proclaim: "Join our side. They are wrong; we know what to do!" What follows these proclamations is generally a mixture of reasonable and sophisticated arguments that offer only tenuous support. This type of discourse leaves many people confused, skeptical, and disengaged. In the words of fifteenth-century skeptic Michel de Montaigne, many people feel as if "all appear right in their turn, even though they contradict each other."[3]

These last comments point to the most obvious problem with adversarial debate: antagonists frequently resort to whatever techniques work to win. For instance, in the cases of both nanotechnology and biotechnology, proponents of these innovations have made unrealistic claims about their potential to solve problems, while opponents have raised alarms that seem disconnected from reality. On one side, these new technologies are advertised as magic bullets for some of the most chronic and troubling problems humanity faces in the areas of health, hunger, and environmental degradation. On the other side, they have been labeled as possible agents of irreversible global catastrophes in these same areas. In addition, some people on both sides of the debate vilify each other with frequent ad hominem attacks. Scientists have been labeled as profiteering, blasphemous tinkerers of nature's sacred essence. Activists have been labeled as antiscience zealots who lack compassion for the world's poor and hungry. Unfortunately, there seems to be little penalty for using unscrupulous argumentative tactics, which are often effective.

Paul Thompson, a leading bioethicist who has followed the agricultural biotechnology debate for over twenty-years, describes this kind of speech as strategic discourse. He defines strategic discourse as communication that seeks to manipulate people's attitudes and behaviors. Thompson then remarks, "I hope it is evident to everyone that strategic discourse is never an appropriate response to an ethical issue."[4] Strategic discourse is obviously unethical in science and technology debates because it involves misinformation, deception, and manipulation in an effort to advance private interests while disregarding the common good.

Finally, the debate format presupposes a zero sum competition: if one side wins, the other loses. This format might be appropriate in academic debates where opposing theses must be argued out and they both cannot be true. However, in cases where people are making collective decisions on alternative courses of action that affect the common good, debate is often inappropriate. Public debates seem to reinforce the perception that groups of people hold irreconcilable values that lead to different goals and competing solutions to problems. However, when it comes to science and technology debates about medical research, global climate change, and the future of agriculture, for example, this is not true. In these situations, if bad decisions are made about large-scale problems, some people could win for a period of time but in the long run, humanity and the environment stand to lose big. In the case of global climate change, how we interpret the science, deal with the risks, and select policies and technologies to respond to the crisis will have far-reaching consequences for the future of civilization and the biosphere. The same is true of the future of agriculture and global health. The planet will soon be home to more than 9 billion people and our decisions about how we grow the food needed to feed everyone, and meet their medical needs, will play a large role in determining the fate of civilization and the biosphere.

Again, all this is not to say that there is no place or time for debate or that deliberation is without conflict and problems. Debate teaches the important intellectual skills of articulating and defending a position and criticizing opposing ones. However, it does not develop the mental habits of systematically and comprehensively working through difficult and uncertain issues with others. Deliberation is the more appropriate form of dialogue when the question, "What should we do?" involves strong elements of risk and serious moral implications.

COHERENT AND SYSTEMATIC DELIBERATIONS

Deliberation involves much more than a group of people getting together to discuss the issues in a spirit of cooperation. There is a discernable logic or pattern to systematic and coherent deliberations.[5] The general structure of deliberative dialogue is to define problems, characterize goals, and list and select the most appropriate means from a range of alternatives to realize those goals. Confusion and disagreement can be avoided by structuring deliberations as a coherent whole that aims towards a conclusion. There are at least six areas of discussion in the deliberative process: (1) Defining problems, (2) characterizing goals, (3) listing competing alternative means to reach those goals, (4) discussing obstacles associated with particular alternatives, (5) analyzing unwanted side effects that might arise with particular alternatives, and (6) articulating a conclusion in the form an imperative for action.

In answering the question: "What should we do?" deliberators begin by defining problems and characterizing and prioritizing goals. This is only common sense. For example, it is important to understand the reasons motivating a trip and what one hopes to accomplish at the destination before deciding the best route and/or means of travel.[6] The activities of defining problems and characterizing goals provide the context for evaluating alternative means to solving a problem and realizing a goal. For example, in the case of agricultural biotechnology, this technology is introduced into a world where agriculture faces severe and far-reaching challenges. The soil scientist Daniel Hillel states the problem succinctly: "At the same time the people of the earth are proliferating, their treatment of the earth is diminishing its capacity to support them."[7] Agricultural yields in many places are stagnant or declining and populations are growing; at the same time high-yield industrial agriculture is responsible for diminishing the earth's ecosystem services.[8] As a consequence, over the next fifty years world agriculture faces the daunting challenges of achieving the interdependent goals of food security for 9 billion people and making agricultural practices ecologically sustainable.

Once the dialogue over goals has proceeded to a high level of sophistication, deliberators ask: What alternatives exist for pursuing these goals? Deliberation is done in light of competing alternatives. The task is to develop a provisional list of viable alternative means for achieving the stated goals. This is not a theoretical exercise, but a pragmatic one. The starting point is real-life decisions, for example, over medical research, biotechnology, nanotechnology, and global climate change. The list of alternatives could theoretically be very long and could easily drift into science fiction and/or political fantasy. The lines between sound science, political reality, and fanciful thinking must be clearly determined.

With goals and alternatives established, the next task is to ask: Are there practical obstacles that make it impossible or improbable to achieve these goals with these means? Can we evaluate means according to their practicality and efficacy? Does the science support the means? Is there sufficient funding available? Are there institutions in place that would allow for the application or diffusion of this technology? Once again, fruitful dialogue on practical obstacles must be interdisciplinary. Deliberators can intelligently discuss these questions only if there is accurate knowledge of the science involved. Further, there must be ethical and political sophistication in discussing questions about how scientific research is funded and the institutional relationships among science, business, and government.

At the same time, deliberators must pursue the question: In pursuing these goals with this means, are we likely to generate unwanted side effects? There are at least three general classes of unwanted side effects that need to be discussed: health, environmental, and socioeconomic. For example, nanotechnologies raise numerous as yet unanswered questions about the biological effects of particles small enough to make their way into the nucleus of the cell or to cross the blood-brain barrier. The debate over agricultural biotechnology has largely focused on unintended health and environmental consequences. Critics of agricultural biotechnology frequently provide "laundry lists" of the potential negative health and environmental consequences of transgenic crops and foods. Also, new technologies are not socially (economically and politically) neutral. Technological change frequently transforms ways of life and creates economic winners and losers.[9] For example, in the nanotechnology debates, there are worries about the creation of a "nanodivide" between those who can afford the technologies and those who cannot. Similar concerns are raised in the climate change and biotechnology debates. In all three areas deliberators need to explore the question: in pursuing these goals with this means, are we likely to create more, or worse, problems than we solve?

In addition, since these questions are often about *potential* side effects, ethical and scientific discussions are needed to determine how to deal with the uncertainty inherent in such predictions. The recent intense discussion over the "precautionary principle" highlights this point. Deliberators must also discuss how to deal with the uncertainty inherent in such predications in a scientifically and ethically responsible manner.

Before moving to a conclusion, and in light of the research and discussions in these areas just listed above, deliberators should discuss the following types of review questions:

- Is the goal adequately understood and agreed upon?
- How should this goal be prioritized in relation to other goals?
- Have all realistic alternatives been listed and adequately evaluated?
- Have all practical obstacles been considered and thoroughly evaluated?
- Have all significant side effects been anticipated and appraised?
- How do we select the best means?
- Is the selected means really the best alternative? Or, should we reexamine the list of alternatives?

Once these review questions have been covered, deliberators are in a good position to recommend a course of action. It is possible that there may not be complete agreement. A number of the key questions (for example, the exact health effects, the possible impacts on dynamic systems, the possible development of future technologies) will be impossible to answer completely due to genuine scientific uncertainty, and differences in values may cause people to see these risks differently. However, after going through this process of practical reasoning and ethical dialogue, deliberators will understand where the remaining uncertainties are likely to be located and the values that need to be discussed. They will be able to better identify the difference between legitimate obstacles that stand in the way of a decision and political obfuscation. They will certainly be in a better position to identify harmful courses of action and the likely consequences of inaction.

While a greater emphasis on deliberation over debate and systematic, coherent deliberations in applied and research ethics is a step in the right direction, it is not enough. There are additional factors at work that can undermine the deliberative process. The quality of deliberations also depends upon the composition and behaviors of the deliberative groups and the ethical character of the deliberators.

GROUP DYNAMICS AND SOCIAL INFLUENCES ON DELIBERATIONS

The view that group deliberations lead to better decisions goes back at least to Aristotle who wrote: "When there are many who contribute to the process of deliberation, each can bring his share of goodness and moral prudence . . . some appreciate one part, and some another, and all together appreciate all."[10] Individuals have limited knowledge, imperfect reasoning abilities, and finite moral perspectives. All of these limitations are particularly acute in the twenty-first century as humanity faces unprecedented collective action problems like global climate change. These critical limitations of the human condition can be expanded in an open and inclusive deliberative process. Cass Sunstein writes:

> The give and take of group discussion might still sift information and perspectives in a way that leads the group to a good solution to a problem, in which the whole is actually *more* than the sum of its parts. . . . The exchange of views leads to a creative answer or

> solution that no member could generate individually. . . . Groups can outperform their best members, in a way that suggests synergy is involved.[11]

In addition to potentially increasing knowledge, expanding perspective, and deepening moral wisdom, the deliberative process tends to make individuals more confident in their opinions and unified in their positions.[12] These tendencies can serve the positive function of helping people to work together toward a common goal. Because deliberating with others increases one's confidence in one's positions, it can legitimize a chosen course of action. And, because it promotes unity of opinion, it can bring people together to work toward a common end or goal.

Ironically, the same features that promote legitimacy and unity can have negative consequences due to certain kinds of group dynamics and social pressures. Getting groups of people together to discuss problems, goals, and possible solutions does not guarantee an intelligent or ethical outcome. In fact, in closed groups of like-minded people, ignorance can spread like a virus and immoral decisions can metastasize. One only needs to think of closed groups of extremists who migrate toward terrorism and violence. Deliberative groups can function well or poorly as a function of group composition and social dynamics. Cass Sunstein has carefully discussed problems with deliberative groups in several books and articles. In what follows, I will briefly mention just two such problems: group polarization and suppressing information.

In contrast to the competitive structure of debate, the cooperative nature of deliberative discourse can ironically exacerbate the problem of group polarization. This mechanism for group polarization tends to happen when a sub-group of people with similar leanings comes together to discuss a controversial issue. When the group composition is ideologically homogenous, for example, discussions do not produce sophisticated, nuanced, and moderate positions but simplistic and extreme positions. One reason for this is that in closed groups of likeminded people there will be an abundance of facts and arguments supporting the group's original biases and a scarcity of information and counterarguments that might challenge or undermine these biases. Stated differently, the scope of discourse is narrow and constricted, as people with contrary views and conflicting information are absent. In addition, when people hear their views affirmed and supported, they tend to become more confident in their positions. In general, this type of group composition leads people to migrate away from the center and toward the poles. Sunstein observes: "The clear lesson is that when a group is highly cohesive, and when its members are closely identified with it, polarization is especially likely—and is especially likely to be large."[13]

The structure of deliberative groups, then, plays a key role in a group's ability to aggregate the information, ideas, and arguments needed to make good judgments. As Aristotle long ago observed, it is the wide range of expertise and the quality of collective wisdom that makes deliberative groups more effective than individuals. As a rule, closed, uniform groups tend to perform poorly, while open, diverse groups perform much better due to the wider range of facts and arguments at their disposal. However, another problem with group deliberations must be overcome for this general rule to apply.

The other danger of group deliberations is described by the familiar term, groupthink. There is a negative tendency for deliberative groups to emphasize shared information and slight information that is held by only one or a few members.[14] So in order to take advantage of the full spectrum of information, ideas, and arguments, a group dynamic should be created that allows people to speak up and say what they think.[15] Sunstein lists several factors working against this. For instance, it seems to be part of human psychology for most people to feel uncomfortable being the outlier or dissenter in a group. Most people feel more comfortable when their views and opinions are in line with group leaders or dominant opinions. It is difficult and risky to go against the leader or majority, even when one thinks or knows that the leader or group is wrong. People tend to like people who agree with them, and like themselves more when others agree with them. Making this problem worse, if the social dynamics are such that people think their statements "will be disliked or ridiculed, they might not speak out, even on questions of fact."[16] Therefore, even diverse groups can fail to aggregate information or arrive at the best course of action when these kinds of social pressures are present. Sunstein summarizes the situation: "When deliberating groups do badly, fear of social sanction is often a major reason. People should be rewarded for telling the truth and moving the group in a direction that turns out to be the right one."[17] However, he remarks, "People are much more willing to say what they know if other dissenters are present and if a principle of equity is widely accepted in the group."[18]

Both of these problems with group deliberations can be overcome. Sunstein writes: "Deliberation has its own internal morality, one that should overcome some of the harmful effects of deliberative processes in the real world. Perhaps deliberation will work well, and produce accurate results, when it follows that internal morality."[19] One way to move our culture toward more ethical deliberations is through study and practice. It is clear that by paying careful attention to group composition and social influence, problems with group deliberations can by and large be avoided. Diverse and open groups where people feel free to speak up have a good chance of performing well at aggregating information and assessing alternative courses of action. As Sunstein notes in the above quote, group deliberations have an internal morality. In the final section I will explore this issue by turning to ethical theory and the character of individual deliberators.

DELIBERATION, PRACTICAL REASON, AND ETHICAL THEORY

As noted earlier, "ethical deliberation" is a way of emphasizing a systematic effort to include a wide range of value considerations in the deliberative process, as opposed to merely practical or instrumental concerns. To review, ethical deliberation is a form of dialogue for addressing contentious issues where difficult decisions must be made with incomplete information and competing moral positions, especially where risk, complexity, and uncertainty are important. Many minds that possess multiple areas of expertise are needed to aggregate the information to make good decisions. However, recalling Aristotle's words, we also need a variety of people to bring their share of "goodness and moral prudence." Successful deliberations need facts, but also individuals who possess practical wisdom, an ethical attribute.

The ethical attributes of a good deliberator are built upon the human capacity for practical reason. Practical reason simply refers to the "human capacity for resolving, through reflection, the question of what one [should] do."[20] The philosopher, Martha Nussbaum writes of practical reason:

> All humans participate (or try to) in the planning and managing of their own lives, asking and answering questions about what is good and how one should live. . . . This general capacity has many concrete forms and is related in complex ways to other capacities, emotional, imaginative, and intellectual. But a being who altogether lacks this would not be likely to be regarded as fully human in any society.[21]

The development of practical reason into the excellence (moral virtue) of practical wisdom (prudence) is accomplished through the study and practice of deliberation. While the term, practical wisdom, and its synonym, prudence, are not an active part of most people's moral vocabulary, we are all familiar with the consequences of its absence. It is easy to think of at least one pain-in-the-neck moralist whose sincere morality lacks the guidance of practical wisdom. This has been an important factor driving recent science debates, where principles can sometimes trump facts. In all three cases, the nanotechnology, climate change, and agricultural biotechnology debates, one can readily identify well-meaning people guided by deeply felt moral convictions who are championing unreal, irrational, and harmful positions. Such single-minded people can hinder productive deliberations. In short, morality without practical wisdom can be pointless, irritating, or dangerous.[22] Practical wisdom is the ability to fruitfully engage in cooperative deliberations with the purpose of selecting intelligent, moral goals, and to discern appropriate and effective means to achieve those goals. Practical wisdom, like all virtues, is an attribute that people can develop. What exactly does this attribute look like?

In a recent book on deliberative democracy, political philosopher Robert Talisse provides a preliminary list and descriptions of the moral excellences of deliberation.[23] His list includes honesty, modesty, charity, and integrity. Excellent deliberators are honest because they are willing to admit that their favored position might turn out, upon closer examination, to be faulty, narrow-minded, or in need of revision. They are willing to consider all the evidence and to give all proposals a fair appraisal before deciding on a policy. Excellent deliberators are modest because even the best-intentioned plans and policies can be ineffective or fail in practice. Modest deliberators understand that political proposals are not ultimate solutions. Hence they are able to admit error and seek correction. Charity means listening to the proposals one does not agree with. This requires rejecting simplistic categories that give rise to prejudices, such as of "pro" and "con," "left" and "right," "conservative" and "liberal." Excellent deliberators see these polarizing ideological categories as obstacles to deliberation. Finally, the deliberator who embodies the virtue of integrity understands that "however divided he and his fellow citizens otherwise may be, they nonetheless are joined in the common and continuing undertaking of self-government."[24] This requires being committed to the ideal of self-governance through the reasonable exchange of ideas. The purpose of focusing on deliberation as a basis for ethics education is to provide awareness of these excellences.

CONCLUSION

As a reminder, the focus of this chapter and book is to provide food for thought on improving the quality of public decisions on science and technology by putting a greater emphasis on ethical deliberation. It is not advocating a wholesale transformation of our culture—that of course would be unrealistic. However, by encouraging a greater emphasis on ethical deliberation in research and applied ethics, an incremental, bottom-up approach is likely to have a positive influence on public discourse on controversial issues in science and technology.

This introduction to ethical deliberation in the context of major controversies in science and technology has pointed to four general areas of study and research: deliberation and educational practices, the structure of systematic and coherent deliberations, the social dynamics of group deliberations, and ethical theory and deliberation. Given the current ethos of adversarial debate in our larger culture it seems wise to place a greater emphasis on ethical deliberation in courses in research and applied ethics. Through research and study into systematic, coherent deliberations, people will be better able to arrange confused and problematic situations into a coherent whole that points toward a well-considered conclusion. By studying group dynamics and social pressures in deliberative groups, people will be better able to organize and structure more successful deliberative groups. Finally, by studying deliberation and ethical theory, it is possible to enhance institutions and foster the development of practical wisdom.

In the coming century, global society faces difficult collective challenges of great magnitude and complexity involving climate change, agriculture, global health, and the environment. Science and technology will play a key role in meeting these challenges. For democratic societies to fully and appropriately utilize science and technology to solve global problems it would be wise to improve the quality of public discourse. The stakes are too high for conflict and manipulative discourse to dominate these decisions.

2. OVERCOMING SCIENTIFIC IDEOLOGY

Bernard Rollin

Despite the fact that the United States has produced some of the most sophisticated science in human history, as well as truly dramatic technology attendant on that science, the American public is far from unequivocally supportive of science. There are many reasons for that lack of support. Perhaps the most patent reason is the appalling scientific illiteracy and rampant anti-intellectualism that is pervasive throughout American history.

We cannot underestimate the degree of scientific illiteracy rampant in the US public and elsewhere. First described with regard to intellectuals in universities by C. P. Snow in the 1950s as "the two cultures in conflict," i.e., science and everything else, there is little reason to believe things have improved. As Keith Black, MD, wrote in the Cedars-Sinai Neurosciences Report:

> America's best and brightest used to go into science and medicine, but no longer. The United States consistently ranks in low compar-

> ison to other developed countries on assessments of scientific literacy. "One half of the American public does not know the earth goes around the sun once a year, and believes that the earliest humans lived at the same time as the dinosaurs." [NSF]. . . . A 1996 National Assessment of Educational Progress survey found that 43 percent of high school seniors did not meet the basic standard for scientific knowledge.[1]

Dr. Jon Miller of Northwestern University, who studies scientific literacy in the United States, affirms that only 20–25 percent of Americans are "scientifically savvy and alert . . . [the rest] don't have a clue." According to Miller, US adults do not know what molecules are, fewer than a third know that DNA is the key to heredity, and only 10 percent know what radiation is. Sixteen percent of high school science teachers are Creationists. Two thirds of the US public wants creation taught along with evolution, according to a 2004 CBS news poll.[2]

This should not surprise us, given Richard Hofstadter's Pulitzer Prize winning 1964 book, *Anti-Intellectualism in American Life,* which points out the deep current of anti-intellectualism in American history going back to the founding of this country. And not only is the United States anti-intellectual; we are openly hostile to science. As Jeffrey Sacks wrote in the *Economist,* September 22, 2008: "By anti-intellectualism I mean an aggressively antiscientific perspective, backed by disdain for those who adhere to science and evidence," and consider that stem cells and biotechnology have been widely rejected for bad reasons.[3]

Other factors both follow from and augment antiscience feeling. These include the unfortunate resurgence of "magic thinking," in turn reflected in the reappearance of Creationism hostile to evolution and the billions of dollars spent on evidentially baseless "alternative medicine"; and the fact that cryptozoology books sell more than all bio-science books. The never-ending appeal of the Frankenstein myth as a metaphor for scientific progress in turn ramifies in public skepticism regarding scientific advances.

In this essay I will discuss a major additional threat to the continued thriving of science in our society, all the more insidious because it is largely unrecognized by those in the scientific community in a position to rectify the problem. The one threat I will detail here has escaped mainstream serious attention.

I will begin by relating a series of fairly shocking anecdotes whose unifying thread I will discuss in due course. I ask you to try and think through that thread before I explain it. I hope to elicit the shock of recognition from you by following Plato's dictum that, when dealing with adults and ethics, one cannot teach, only remind:

1) In about 1990, the then Director of NIH, and therefore arguably the chief representative of biomedicine in the United States, was visiting his alma mater. He was talking to a group of students informally and was apparently unguarded in his remarks, not realizing that a student reporter for the school paper was present. The students asked him about the ethical issues associated with genetic engineering. His reply was astonishing: He opined, "Though scientific advances like genetic engineering are always controversial, science should never be hampered by ethical considerations."
2) Around 1980, when I was developing and pressing the federal legislation for laboratory animals adopted in 1985, I was invited by the American Association of Laboratory Animal Science (AALAS) to discuss my reasons for supporting legislative constraints on science on a panel with half a dozen eminent laboratory animal veterinarians. By way of making my point, I asked them all to tell me what analgesic would be of choice for a rat used in a limb-crush experiment, assuming analgesia did not disrupt results that were being studied. The consensus response was, in essence, "How should we know? We don't even know for sure if animals feel pain!" Interestingly, five years later, after the laws had been passed, I phoned one of those veterinarians, pointing out that as of now, he was *required* to know the answer to my question. He then rattled off five different analgesic regimens. I asked him, "You were agnostic five years ago, where did you get your information?" "From the drug companies!"

he said. Puzzled, I asked if they now worked on rat analgesia. "No," he said, "but all human analgesics are tested on rats." The point is that he knew this five years earlier but did not then see it as relevant to rat pain!

3) At another AALAS meeting in the early 1980s, I ran a full day session on ethics and animal research. At the end, the reporters present converged on the president of AALAS, asking him to respond to my points. "Oh there are no issues in animal research," he said, "God said we can do whatever we want with animals." (When the reporters asked me to respond, I facetiously said that what he said could not possibly be true. "Why?" they asked. "Because he is at the Blank Vet School," I replied, "and if God chose to reveal himself at a vet school, it would surely be Colorado, which is, after all God's Country.")
4) At the American Veterinary Medical Association (AVMA) pain panel convened in 1986 by Dean Hiram Kitchen at the request of Congress, I was asked to write the prologue to the report. I did, and presented it to the group. I approvingly pointed out that according to the great *skeptical* philosopher, David Hume, few things are as obvious as the fact that animals have thoughts and feelings, and that this point does not escape "even the most stupid." A representative from the National Institute of Mental Health (NIMH) stood up indignantly and declared, "If we are going to talk mysticism, I am leaving," and did, never to return.
5) In the late 1970s, virtually every veterinary school taught surgery by doing multiple survival procedures. Animals were used for practice surgery a minimum of eight times over three weeks at Colorado State University, to over twenty times at some institutions. This was done to save money, and the ethical issues occasioned were never discussed, nor did the students dare to raise them until my 1978 ethics class at Colorado State University, which undid this odious practice. In 1980, when I was well into teaching the ethics and animal welfare course at the veterinary school, I learned that the first laboratory exercise required of the students in the *third week* of their first year was

to feed cream to a cat and then, using ketamine (which is not an effective analgesic for visceral pain but instead serves to restrain the animal), do exploratory abdominal surgery, ostensibly to see the transport of the cream through the intestinal villi. When I asked the teacher what was the point of this horrifying experience (the animals vocalized and showed other signs of pain), he told me that it was designed to "teach the students that they are in veterinary school, and needed to be tough, and that if they were 'soft,' to 'get the hell out early.'"

As late as the mid-1980s, and even into the current century, many veterinary and human medical schools *required* that the students participate in bleeding out a dog until it died of hemorrhagic shock. Although Colorado State University's veterinary school abolished the lab in the early 1980s for ethical reasons, the department head who abolished the lab was defending the same practice ten years later, after moving to another university, and explained to me that if he didn't, his faculty would force him out. As late as the mid-1990s, a medical school associate dean told my veterinary dean that his faculty was "firmly convinced" that one could not "be a good physician unless one first killed a dog." In his autobiographical book, *Gentle Vengeance,* which deals with an older student going through Harvard Medical School, the author remarks in passing that the only purpose he and his peers could see to the dog labs was to assure the students' divestiture of any shred of compassion that might have survived their premedical studies.

At one veterinary school, a senior elective class provided each student with a dog, and the student was required to do a whole semester of surgery on the animal. One student anesthetized the animal, beat on it randomly with a sledgehammer, and spent the semester repairing the damage. He received an A.

6) In 2001, I was part of the World Health Organization group that was charged with setting guidelines for the use of antibiotics in animal feeds, since their indiscriminate use was

driving evolution of resistance to antimicrobials and endangering human health. I was asked to give the keynote speech defining the ethical dimensions of the issue. When I finished, I asked for questions. One veterinarian who was in fact the Food and Drug Administration employee in charge of managing the issue leapt up and said to me "I am offended!" "By what?" I asked. "By the presence of an ethics talk at a scientific conference. Ethics has nothing to do with this issue! It is strictly a scientific question!" Stifling an urge to strangle her, I calmly said, "Let me show you that you are wrong. Suppose I give you an unlimited research budget to determine when to stop or curtail the use of antibiotics in feeds. We do the research and find out that current use levels kill (or sicken) 1 person in 500 or 5,000 or 50,000 or 500,000 or 5,000,000. Even when we know this data, it does not tell us where the risk of morbidity or mortality tells us to discontinue such antibiotic use. That is an ethical decision!" She then kept quiet for the rest of the conference!

7) In the early 1980s, when my colleagues and I had pretty well drafted the key concepts of our proposed federal laboratory animal legislation and Colorado Representative Pat Schroeder had committed to carrying it forward, we were told by congressional aides that we needed to provide clear evidence for the need for such law, both because the medical research community was a major financial contributor to congressional campaign war chests and because that same community claimed to be already controlling pain in research animals. In essence, I was charged with proving they were not. I did so by doing a literature search on analgesia for laboratory animals or, indeed, any animals. What did I find? Two references: one of which said there ought to be papers on this issue, and one saying we don't know much, but here is the little we know—aspirin, morphine, and that was about it. If analgesia were indeed widely used, I told Congress, I would have been able to find a significant literature on its theory and practice.

8) In 1979, I attended a conference on animal pain, where I debated a prominent scientist, with me defending the view that animals could feel pain while he denied that claim. I thought we had enjoyed an amicable discussion until I returned to Colorado State University, where I found out that after the debate he had called the dean of veterinary medicine and told him that I was "a viper in the bosom of biomedicine" who should not be allowed to teach in a veterinary program!

   The first text books of veterinary anesthesia published in the United States, by Lumb[4] and by Lumb and Jones,[5] do not even mention felt pain, even as a reason for anesthesia, while they do list positioning the limbs, avoiding getting hurt by the animal, etc. And the American Physiological Society volume of 1982, in part designed to assure the public of science's concern with animal pain, and titled *Animal Pain: Perception and Alleviation*,[6] does not talk about perception or alleviation, but focuses on the "plumbing"—electrochemistry and physiology—of pain, with only one small reference to its alleged purpose, in a paper by Lloyd Davis.[7]

9) As the Talmud says, "the last is the most beloved," so I will conclude with one of my most loved anecdotes. A few years before Dolly the cloned sheep was announced, I received a Saturday afternoon call from a research official at the Roslin Institute, asking me to chat about the ethics of "hypothetically" producing a cloned animal. "It's your nickel," I said, finally deploying that wonderful locution stolen from old "hard-boiled" 1940s detective novels and noir films, "Keep talking!" I told them that there were two major concerns: does cloning harm the animal, and does it create any social, ecological, or disease dangers. More important than these legitimate concerns, I continued, was what I called "A Gresham's Law for Ethics."[8] Recall that Gresham's Law in economics asserts, "Bad money drives good money out of circulation." In the same way, "Bad Ethics drives good ethics out of circulation." So, for example, after World War I, German currency (the Deutschemark) was so inflated that it took a

> wheelbarrow full of them to by a loaf of bread. In such an economy, rational people pay their outstanding debts with Deutschemarks, not with gold, which they hoard. So too in ethics, I continued. Any new technology, be it the computer or biotechnology, creates a vacuum in social ethical thought, and fear. "What effects will this have on our lives? Is it good or bad? What do we need to control?" If the scientists do not inaugurate rational discussion, doomsayers with vested interest, such as Jeremy Rifkin, will fill that lacuna. "So," I concluded, "that is your biggest worry. You must create an educated populace on cloning, and help them define the issues, or the public will be told that it 'violates God's will,' and how can you respond to that?"
>
> As I suspected, it was as if our conversation had never taken place! Some years later, Dolly was announced to a completely uninformed public. Time/Warner did a survey one week after the announcement. Fully 75 percent of the US public affirmed that cloning "violated God's will!"

I have many other similar stories; for example, I could have discussed the period where open-heart surgery was done on neonatal humans without anesthesia using curariform drugs; or I could have discussed how the primatology branch of the American Psychological Association, when approached by Animal and Plant Health Inspection Service (APHIS) chief Robert Rissler after the 1985 laboratory animal laws passed to help define "psychological well-being of primates," replied, "Don't worry there is no such thing," and Rissler classically rejoined, "There will be after January 1, 1986, whether you people help me or not." But I have said enough, I hope, to establish our universe of discourse.

All of these anecdotes illustrate the prevalence of what I have called Scientific Ideology, or alternatively, the Common Sense of Science, for it is as ubiquitous in science as ordinary common sense is in daily life.

Ideologies operate in many different areas—religious, political, sociological, economic, ethnic. Thus it is not surprising that an ideology would emerge with regard to science, which is, after all, the dominant way of knowing about the world in Western societies since the Renaissance.

Indeed, knowing has had a special place in the world since antiquity. Among the pre-Socratics—or *physikoi* as Aristotle called them—one sometimes needed to subordinate one's life unquestioningly to the precepts of a society of knowers, as was the case with the Pythagoreans. And the very first line of Aristotle's *Metaphysics*—or First Philosophy—is "All men by nature desire to know." Thus the very *telos* of humanity, the "humanness" of humans, consists in exercising the cognitive functions that separate humans from all creation. Inevitably, the great knowers, such as Aristotle, Bacon, Newton, and Einstein, felt it necessary to articulate what separated legitimate empirical knowledge from spurious knowledge to guard and defend that methodology from encroachment by false pretenders to knowledge.

Thus the ideology underlying modern (i.e., post-medieval) science has grown and evolved along with science itself. And a major—perhaps *the* major—component of that ideology is a strong positivistic tendency, still regnant today, of believing that real science must be based in—and only in— experience, since the tribunal of experience is the objective, universal judge of what is really happening in the world.

If one asks most working scientists what separates science from religion, speculative metaphysics, or shamanistic worldviews, they would unhesitatingly reply that it is an emphasis on validating all claims through sense experience, observation, or experimental manipulation. This component of scientific ideology can be traced directly back to Newton, who proclaimed that he did not "feign hypotheses" ("*hypotheses non fingo*") but operated directly from experience. (The fact that Newton in fact *did* operate with nonobservable notions such as gravity or, more generally, action at a distance, and absolute space and time, did not stop him from issuing an ideological proclamation affirming that one should not do so.) The Royal Society members apparently took him literally, went around gathering data for their commonplace books, and fully expected major scientific breakthroughs to emerge therefrom.

The insistence on experience as the bedrock for science continues from Newton to the twentieth century, where it reaches its most philosophical articulation in the reductive movement known as *logical positivism,* a movement that was designed to excise the unverifiable from science and, in some of its forms, to axiomatize science so that its derivation from observations was transparent. A classic and profound example of the purpose of the excisive dimension of positivism can be found in Einstein's rejection of Newton's concepts of absolute space and time on the grounds that such talk was untestable. Other examples of positivist targets were Bergson's (and other biologist's) talk of life force (*élan vital*) as separating the living from the nonliving, or the embryologist Driesch's postulation of "entelechies" to explain regeneration in starfish.

Although logical positivism took many subtly different and variegated forms, the message, as received by working scientists, and passed on to students (including myself), was that proper science ought not allow unverifiable statements. This was no doubt potentiated by the fact that one British philosopher, a logical positivist named A. J. Ayer, wrote a book relatively readable, vastly popular (for a philosophy book), and aggressively polemical, that defended logical positivism, entitled *Language, Truth, and Logic;* it first appeared in 1936 and has remained in print ever since.[9] Easy to read, highly critical of wool-gathering, speculative metaphysics, and other soft and ungrounded ways of knowing, the book was long used in introductory philosophy courses and, in many cases, represented the only contact with philosophy that aspiring young scientists—or even senior scientists—enjoyed.

Be that as it may, the positivist demand for empirical verification of all meaningful claims became a mainstay of scientific ideology from the time of Einstein to the present. Insofar as scientists thought at all in philosophical terms about what they were doing, they embraced the simple, but to them satisfying, positivism we have described. Through it, one could clearly, in good conscience, dismiss religious claims, metaphysical claims, or other speculative assertions not merely as false, and irrelevant to science, but as meaningless. Only what could be verified (or falsified) empirically was meaningful. "In principle" meant "someday," given technological progress. Thus though the statement "there are intelligent inhabitants on Mars" could not in fact be verified or falsified in 1940, it was still meaningful, since we could see how it could be verified, i.e., by building rocket ships and going to Mars to look. Such a statement stands in sharp contradistinction to the statement "there are intelligent beings in heaven" because, however our technology is perfected, we don't even know what it would be like to visit heaven, it not being a physical place.

What does all this have to do with ethics? Quite a bit, it turns out. The philosopher Ludwig Wittgenstein, who greatly influenced the logical positivists, once remarked that, if you take an inventory of all the *facts* in the universe, you will not find it a *fact* that killing is wrong.[10] In other words, ethics is not part of the furniture of the scientific universe. You cannot, in principle, test the proposition that "killing is wrong." It can neither be verified nor falsified. So, in Wittgenstein's view, empirically and scientifically, ethical judgments are meaningless. From this, it was concluded that ethics is outside of the scope of science, as are all judgments regarding values rather than facts. The slogan that I learned in my science courses in the 1960s, and which has persisted to the present and is still being taught in too many places, is that "science is value-free" in general, and "ethics-free" in particular.

The denial in particular of the relevance of ethics to science was taught both explicitly and implicitly. One could find it explicitly stated in science textbooks. For example, in the late 1980s, when I was researching a book on animal pain, I looked at basic biology texts, two of which a colleague and I actually used, ironically enough, in an honors biology course we team-taught for twenty-five years, which attempted to combine biology and the philosophical and ethical issues it presupposed and gave rise to. The widely used Keeton and Gould textbook,[11] for example, in what one of my colleagues calls the "throat-clearing introduction," wherein the authors pay lip service to scientific method, a bit of history, and other "soft" issues before getting down to the parts of a cell and the Krebs cycle, loudly declares that "science cannot make value judgments . . . cannot make moral judgments."[12] In the same vein, Mader, in her popular biology text asserted, "Science does not make ethical or moral decisions."[13] The standard line affirms that science at most provides society with *facts* relevant to making moral decisions, but never itself makes such decisions.

We have argued that the logical positivism that informed scientific ideology's rejection of the legitimacy of ethics dismissed moral discussion as empirically meaningless. That is not, however, the whole story. Positivist thinkers felt compelled to explain why intelligent people continued to make moral judgments and continued to argue about them. They explained the former by saying that when people make assertions such as "killing is wrong," which seem to be statements about reality, they are in fact describing nothing. Rather, they are "emoting," expressing their own revulsion at killing. "Killing is wrong" really *expresses* "Killing, yuk!" rather than describing some state of affairs. And when we seem to debate about killing, we are not really arguing ethics (which one can't do any more than you and I can debate whether we like or don't like pepperoni), but rather disputing each other's facts. So a debate over the alleged morality of capital punishment is my expressing revulsion at capital punishment while you express approval, and any debate we can engender is over such factual questions as whether or not capital punishment serves as a deterrent against murder.

It is therefore not surprising that when scientists were drawn into social discussions of ethical issues they were every bit as emotional as their untutored opponents. It is because their ideology dictates that these issues *are nothing but emotional*, that the notion of rational ethics is an oxymoron, and that he who generates the most effective emotional response "wins." So, for example, during the debate on the morality of animal research in the 1970s and 1980s, scientists either totally ignored the issue or countered criticisms with emotional appeals to the health of children. For example, in one film, *Will I Be All Right, Doctor?* (the questions asked by a frightened child of a pediatrician), made by defenders of unrestricted research, the response was "Yes, if *they* leave us alone to do what we want with animals." So appallingly and unabashedly emotional and mawkish was the film that when it was premiered at the American Association for Laboratory Animal Science meetings at a subsection of laboratory animal veterinarians, a putatively sympathetic audience, the only comment forthcoming from the audience came from a veterinarian who affirmed that he was "ashamed to be associated with a film that is pitched lower than the worst antivivisectionist clap-trap!"

Other ads placed by the research community affirmed that 90 percent of the animals used in research were mice and rats, animals "people kill in their kitchens anyway." Sometimes questions raised about animal use, as once occurred in a science editorial, elicited the reply that "animal use is not an ethical question—it is a scientific necessity," as if it cannot be, and is not, both.

My thesis, then, is that an ideologically ubiquitous denial of the relevance of values in general and ethics in particular to science created blinders among scientists to issues of major concern to society. But that is not all. There is another major component of scientific ideology that harmonized perfectly with the value-free dictum. That was the positivistic/behavioristic thesis that science could not legitimately talk about consciousness or subjective experiences, which led to a question about their existence. (John Watson, the founder of Behaviorism came close to saying that we don't have thoughts, we only think we do!) The two mutually reinforcing components of scientific ideology taken together have caused incalculable damage to science, society, and objects of moral concern in science's failure to engage ethical issues. We have exemplified earlier some of the ways that animals suffered.

This second component of scientific ideology affirming that scientists needed to be agnostic about subjective states in people and animals led to the denial of the scientific reality of pain that in turn ramified in open heart surgery performed on neonatal humans until the late 1980s using paralytic drugs, not anesthetics; failure to provide adequate analgesia to all human patients, as recounted in Marks and Sacher's famous paper;[14] and failure to control any animal pain in research at the expense of both animals and science. A healthy dose of philosophy of mind (as well as legislation) was needed to even begin to correct this. This hurt not only innocent creatures, but also science itself by a virtually total failure to control pain and stress variables. All of this ramified in the practice of both human and veterinary medicine in major neglect of pain control. (I have devoted an entire chapter to this neglect in my book *Science and Ethics*.[15])

In the area of human research, the abuses of humans have been legion, from the Tuskegee studies to the recent death of Jesse Gelsinger to the radiation studies performed by the Department of Energy to the release of microbes in subway systems to Willowbrook, etc. And the results for science have been equally pernicious—the Federal imposition of draconian rules for researchers; rules that change with political correctness (allowing pregnant women to serve as research subjects; a flurry about what to call Indians when they are the subjects of research [the government insisted on "Native Americans" until they actually *asked* the Indians]; asking that alcoholics be on committees evaluating study of alcoholics). The net effect is that researchers see these rules as bureaucratic hoops to jump through with absolutely no grasp of the ethical issues, further creating hostility to doing human research properly!

It is my contention that a good part of the reason that society has moved from 1950s optimistic and naïve Buck Rogers, Jetsonion adoration of a science-shaped future to today's wholesale skepticism about science, evidenced in movements like alternative medicine, neofundamentalist anti-Darwinism, rejection of stem cell technology (especially in Europe), and large-scale rejection of biotechnology, is the Greshlam's Law for Ethics I described earlier. Absent scientists responsibly articulating the dangers and moral issues raised by science, opportunists such as theologians, Rifkins, and Luddites will leap into the breach and fill the gap with lurid, uninformed, but highly marketable pseudo-ethical issues, as we saw in the Dolly case. Unless science begins to engage ethics in serious ways, the forces of darkness and unreason and anti-intellectualism will prevail, which are in any case lurking not terribly far from the surface in society. George Gaskell has revealingly shown that Europeans reject biotechnological modalities for reasons of moral concern, not out of fear, as is widely believed.[16]

As I see it, there is only one way to resolve this malignancy; through education of a new generation of scientists trained to think of ethics and science together. Only education can displace, uproot, and supplant ideology. When I worked to draft the federal laws for laboratory animals, it was not to create a regulatory bureaucracy, or, as I put it then, to put "a cop in every lab." I personally tend towards resistance to laws; I will not ride my Harley-Davidson in states that have mandatory helmet laws. But I was enough of a 1960s product to realize the lesson I learned from Martin Luther King and Lyndon Johnson: just because we may have too many laws in some areas does not mean we have enough in others! We drafted our laws in the way Wittgenstein saw his philosophy: as a ladder to reach a higher plane, which later can be thrown away. Just as King envisioned a generation possessed of racial tolerance not needing marshals or paratroopers to escort children to school, so I saw the laws as forcing animal researchers to reflect in ethical terms, so eventually the laws would not be needed. And it worked: from two meager papers on animal pain I found in 1982, the literature has grown to over eleven thousand in twenty-five years! And of course institutional animal care and use committees routinely discuss ethical issues. And this ideology is nowhere near as ubiquitous as it was twenty years ago.

But recall, and note well, that a different bill was being pressed by activists, the "Research Modernization Act," which would have cut the federal research budget by up to 60 percent and put the money into "alternatives." What did "alternatives" mean? As the chief proponent once said, "Oh, you know; plastic dogs that howl when cut and bleed ketchup so they can do their experiments." This is the potential price of ignoring ethics.

The National Institutes of Health (NIH) seems to have somewhat grasped the need for ethics training when it mandated courses in science and ethics for institutions where students received training grants. As I understand it, they were prompted to do this by increasing numbers of reports of data falsification, plagiarism, and what is generically called "misconduct in science." As I pointed out to some NIH officials in the early 1980s, "You can't teach students that science is value-free in general and ethics-free in particular, and then fault them for not behaving ethically!"

I have been teaching science and ethics in some form steadily since 1976. For twenty plus years, as mentioned earlier, I taught a year-long, honors biology course with a biologist, where ethical and conceptual issues arising in science were taught as integrally related to the science we taught, as cake, not icing. Our course was wildly successful; to this day, my coteacher and I still receive communications from former students who are researchers, physicians, and faculty members to thank us for opening their eyes. I get the same response from the students I teach in the NIH graduate course and other graduate courses, including some positive responses from faculty members!

Although I was often viciously accused of being antiscience when I championed the cause of laboratory animals, nothing could be further from the truth. How on earth could I, who love learning above all else besides family, oppose the best route to knowledge ever devised? I used to be called antiscience when I discussed ethical issues in animal research. Ironically, today I am viciously attacked by others, even by my veterinary colleagues, as being too proscience because I coauthored a book attacking alternative medicine as not standing on a solid basis of scientifically garnered evidence.

In any case, the only way to escape the quicksand into which support for science is sinking is to create educational requirements for every graduate student and undergraduate majoring in any area of science. (Too many undergraduates are still taught the mantra, "Science is value-free.") We can rest only when examination of ethical issues presuppositional to and generated by science are as much second nature to scientists as are the double helix or the Krebs cycle and we fully expunge the pernicious ideology we have described. Only in this way can we assure an academic universe congenial to serious attention to ethics occasioned by science. Increasingly Draconian regulation of the sort we have seen in the area of research on human subjects can only alienate researchers. Especially for thinking people, ethics is far more valuable than regulation.

3.

OVERCOMING PHILOSOPHOBIA

A Few Ethical Tools for the Science Debates

Christopher J. Preston

Science and ethics have long occupied different portions of the intellectual spectrum. For numerous reasons—some good, some bad—graduate students in science and engineering tend to have little training in ethics beyond general "standards of practice" rules against moral transgressions such as plagiarism and falsifying data. Even though philosophical ethics is certainly not relevant to every facet of scientific research, it has deep significance to some.

Deliberations over the direction of science and technology research, for example, have clear ethical dimensions. Such research often has a direct effect on the lives and futures of large numbers of people, bringing ethical concerns immediately to the forefront. Some think that science and technology policy should proceed democratically according to choices that are consistent with the public will, others think the agenda should be set by the curiosity of the scientist alone. Questions about justice and fairness are at the very center of deliberations over whether to put research money into appropriate technologies such as small-scale photovoltaic applications, into previously out-of-favor options such as nuclear power plants, or into large-scale geoengineering projects such as the injection of stratospheric aerosols. The products of scientific research often embody particular value sets and commitments, ensuring that scientists are always shaping the moral landscape in their work, whether they are conscious of it or not. After scientific goals have been set, it seems equally clear that research itself ought to proceed along pathways that are fair, safe, prudent, realistic, and economical. While the work of the scientist is focused primarily on investigating "just the facts," this work proceeds in a medium that is always thoroughly saturated with ethical considerations. As a result, a basic familiarity with some key ethical concepts and ideas is a useful addition to the skill set of any scientist.

THE PROBLEM OF PHILOSOPHOBIA

Several years ago, I was attending one of those brown bag lunch events that take place on university campuses where faculty and graduate students from different disciplines get together to learn about and discuss some issue of common interest. The topic that day was coastal water quality. We were told about a particular impoverished community in the region struggling to balance its economic development interests with the health of the local aquatic ecosystem. During the discussion portion of the hour, a senior marine science professor took it upon himself to lay out the nub of the balancing dilemma and describe in exquisite detail—and with great elegance—just how difficult the situation for the community had become. Having done a masterful job of laying out the problem and showing exactly where the difficult decision lay, the senior professor then turned to me (the only philosopher in the room) and said, "Now we need the ethicist to step in and tell us what to do."

The professor's comments illustrate a pervasive belief in interdisciplinary settings, a belief that appears to be particularly prevalent among scientists. This is the belief that the job of solving difficult ethical problems lies solely in the hands of ethicists. The prevailing view is that, whereas it is the job of scientists to solve scientific problems, it is the job of ethicists to solve ethical problems. Take any challenging dilemma facing society, the ethicist, it is thought, has at hand a toolbox that allows them to resolve the problem in the way that a mathematician solves a differential equation or a car mechanic solves the problem of leaking brake fluid. While struggling to formulate an appropriate response to the marine scientist's question, I started trying to figure why this mistaken belief is so common.

The answer seems to be in part related to academic specialization. Scientists, it is thought, have one particular role, while ethicists, historians, poets, and writers each have another. Due to important considerations related to the objectivity of their work, scientists, it is further supposed, are more or less *prohibited* from engaging tough social and ethical questions concerning what should or should not be done. Science is about 'just the facts' and so any subjective proselytizing on values and preferences should be left to somebody else. This positivistic culture of science education has consequently steered scientists away from anything but the most superficial engagement with philosophical ethics.

While this is not the venue to argue the point in detail, it is clear that the idea that scientists can entirely hand ethical work over to others and opt out of the ethical discussion is mistaken for numerous reasons. In the first place, scientists are the people best informed about the reality of both the problems and the technologies that might be used to fix them. More than anyone else, the scientist knows the scientific and technical hurdles that need to be overcome before certain problems can be remedied. Second, scientists are not just experts in research, they are also *citizens* who typically live in the society that will receive the fruits (or bear the costs) of decisions made about the direction of research. Third, many scientists are often supported by public research money or teach at public institutions. It is arguable, then, that they owe the public something in return for the tax money that funds their research and their salaries. For these and other reasons, there is a strong presumption that scientists and scientists-in-training should not pretend to wipe their hands clean of difficult ethical decision-making.[1]

Unfortunately, scientists willing to enter these discussions quickly find themselves in difficult territory. Academic campuses are often divided both intellectually and physically, with the sciences and the humanities and/or social sciences occupying different portions of the landscape. Due to the unrelenting demands of specialization, efforts at interdisciplinarity often pay only lip service to the idea of genuine training in a different discipline. A bio-engineer hoping for a career in the use of nanotechnology in drug delivery, for example, is unlikely to be encouraged to take a class that contrasts Kantian and Utilitarian reasoning. Graduate students often have to brave advisors with a different idea of what their curriculum should look like. The fixed number of hours in a day sets limits on what it is possible for a student to do.

For those lucky enough to find a way around the institutional obstacles, a further barrier presents itself, the psychological barrier of "philosophobia." In the minds of many academics, a large "no trespassing" sign appears at the entrance to another discipline along with a creeping sense of inadequacy concerning any attempt to discuss topics that lie in a different specialty. Few people, after all, have a PhD in more than one discipline. Experts within each discipline have a tendency to make sure that the "no trespassing" sign remains in good repair, often by developing languages and methodologies that require fairly extensive training before anyone from outside can participate. Philosophy in particular has been very good at this gatekeeping, often putting out obscure arguments in impenetrable texts that reinforce the gap between it and other disciplines. As a result, there is an alarming prevalence across academic campuses of philosophobia, as evidenced by the marine science professor mentioned above. Philosophobia can be particularly acute among young scientists, an impending disaster when complex challenges such as those raised by climate change, biotechnology, and nanotechnology need to be carefully thought through. Overcoming philosophobia, then, is a worthy goal for a scientist with a genuine interest in helping to address the pressing issues presented by these and other emerging technologies.

As is the case with many phobias, the treatment of choice is to offer some sort of limited exposure to the source of the fear in a nonthreatening environment, allowing the anxious person to begin to feel more at ease. This is the purpose of the pages that follow.

THE TREATMENT

Philosophical ethics in the Western tradition has tended to be dominated by three main variations of thought. In order not to start with the rather intimidating labels normally used to describe these traditions, let us characterize them initially as: (1) "You must (or must not) do this!" (2) "Weigh the pluses and minuses!" and (3) "Show the virtues of a good deliberator!" All three look like the sort of sound ethical advice that might issue from a morally wise person. Hidden within these statements, however, are three very different approaches to ethical deliberation. If you scratch the surface, these approaches embody three entirely different orientations toward moral goodness.

The first approach simply asserts a rule about a particular behavior that ought to be obeyed. These actions are right; those are wrong. You are obliged to do the former and prohibited from doing the latter. The second invites an opportunity to think about what would follow from adopting a particular behavior. There are upsides and downsides to the proposed course of action. The moral task is to figure out the best balance of good and bad consequences and to act accordingly. The third approach is less focused on the behavior or action itself, but directs attention toward the character of the person making the decision. Is this the sort of thing a morally praiseworthy person would do? Is the decision-maker being just? Is he or she approaching the problem with an open-minded attitude? While the labels used above are an effort to give these approaches a user-friendly appearance, the philosophical names for these three approaches are Deontological, Consequentialist, and Virtue Ethics respectively.

a) Deontology

Deontological approaches to ethics have a long history in the Western tradition, stretching back to Plato, continuing through Christian ethics, and morphing into contemporary rights-based theories. The etymology of the word 'deontology' points toward the idea of a "science of duty." A deontologist studies the reasoning behind why a person might find herself with a duty to act (or refrain from acting) in a certain way. Deontologists look for clear-cut principles that serve as reliable guides to behavior.

What all deontological approaches have in common is the idea that some behaviors are required (or prohibited) on the basis of something inherent in the act itself (or in the intention to act). For a deontologist, there must be something in the nature of the agent's thinking or something in the nature of the victim's treatment that makes the action wrong. Very often it is a matter of the principles that are being obeyed or transgressed. Take, for example, the wrongness of lying. A deontological reason to prohibit lying is that lying *uses the person* lied to in order to further one's own ends. Using someone, a deontologist might suggest, is wrong on principle. In this orientation to ethics, the rightness or wrongness of the action is basic, taking precedence over whatever good or bad consequences might follow from pursuing the action. Even if it appears that some good things might come out of a particular lie, lying remains prohibited on principle. Deontologists are for this reason sometimes described as believing that "the right" (i.e., right action) comes before "the good" (i.e., good results).

Deontological approaches can often be identified by the language used by their practitioners. Deontological language typically has an absolutist tenor, suggesting that these rules come without exceptions and must be obeyed. The injunction to "respect life!" is deontological. "Preserve natural values!" is a command that fits the deontological mold. Appeals to the rules (or principles) established by a divinity are also forms of deontological thinking. "Tampering with genomes means playing God" is a deontological warning against biotechnology. The language of rights is similarly a version of deontology. "The polar bear has a right to arctic ice" would be an example. Deontological prohibitions against actions always point to some important moral law being transgressed. This is why the urgings of a deontologist usually take the form of the statement of certain principles (or "norms") that we are duty-bound to obey.[2]

It is quite natural to suppose that deontology lies at the very center of the study of ethics. After all, ethics is presumably about what one *must* and *must not* do. Deontology, more than any of the other major ethical theories, has the benefit of appearing to supply clear rules for action. Some actions are allowed and encouraged, others are prohibited and discouraged. These encouragements and prohibitions are normally thought to be exportable to all situations. If the rightness or wrongness of a certain action is inherent in the act, then it is unlikely that the particular situation in which the moral decision must be made will affect its morality. Lying will be wrong wherever it takes place.[3] If you believe that tampering with genomes is inherently wrong then it makes no difference if the genome tampering takes place in Western Europe or Sub-Saharan Africa.

The nature of deontology is such that deontologists can sometimes be somewhat uncompromising in their claims, insisting that right and wrong actions are right and wrong in all places and at all times and should be followed or avoided without any qualification. A deontological position can be presented as a moral "line in the sand." The most famous deontologist of all, a Prussian philosopher named Immanuel Kant (1724–1804), is often held up as an example of such an uncompromising thinker, insisting that moral principles were "categorical imperatives," imperatives that must be obeyed at all times, independent of circumstances.

In sum, deontological approaches to ethics have the advantage of apparently offering the clearest guidelines for how to behave, supporting strong cross-cultural intuitions about certain behaviors being categorically wrong, and cohering with a very fundamental intuition concerning what morality is all about. Some of the disadvantages include that they do not make exceptions for circumstances and it can be difficult to know how to resolve conflicts when different principles appear to be at odds with each other.

b) Consequentialism

The second of the major moral theories is often held up as a foil for deontology because its central idea appears to stand the deontological ordering of things on its head. Rather than subscribing to the idea of "the right" coming before "the good," the key idea in this second approach is that one should worry less about the principles involved in particular actions and more about the desirability of the outcome. Like deontology, the approach considers the rightness and wrongness of particular actions as central to morality but the measure of rightness is very different. The approach considers the pluses and minuses of a particular action not in terms of the principles that ground the act (or intention to act) but in terms of whether the action will help to create states of affairs that are considered good. It is because the focus is placed on consequences of action that the approach is referred to as consequentialism.

Like deontology, consequentialism has been a consistent presence since the beginning of ethics, appearing in notable writings from Epicurus to David Hume to John Stuart Mill to Peter Singer. In one of the central versions of consequentialism, the desirable state of affairs being sought is the happiness of the people that will be affected by the action. This most common version, maximizing happiness, is still known as utilitarianism, since the term "utility" is considered a surrogate for "happiness." The call to "maximize happiness" is the most common refrain uttered by consequentialists but it is not the only one. Consequentialists can also seek to 'satisfy the most preferences' or, more generally, to "maximize the number of valuable states" with the discussion of what counts as a valuable state depending on the particular author.[4] This latter form of consequentialism might include creating more objects of beauty, more opportunities for human freedom, or perhaps simply more economic wealth. In all these cases, despite the fact that the measure of the good is changing, the consequences of the act remain the fundamental metric of the act's morality.

The emphasis on results or consequences means that consequentialism is not as strictly focused on adherence to principle as deontology. Obviously the theory includes the principle of always seeking the most desirable results. However, beyond that central principle, hard and fast rules for particular behaviors are of secondary importance. As long as the results are desirable, the method of achieving them is not so important. For this reason, it is often said that in consequentialism "the end justifies the means." Consequentialist reasoning was used to rationalize the dropping of the atomic bombs on Hiroshima and Nagasaki at the end of World War II. The idea that the end justifies the means opens consequentialists up to the accusation of permitting (or even encouraging) barbaric acts to create good consequences. If torturing an innocent civilian maximizes the happiness of the larger population then consequentialism can sometimes seem to favor the torture of innocents. Serious consequentialists have to either bite the bullet on this implication of their theory or find a way to deflect it.[5]

The language of consequentialism can take many forms but it is generally recognizable by the emphasis it puts on what will result from an action. Consequentialist thinking is extremely common in the public arena. All decision-making on the basis of future costs and benefits is consequentialist. Environmental action to avoid harmful ecological effects is consequentialist, as is social policy aimed at increasing the welfare of a nation's citizens. Deciding to act a certain way because that way promises to be 'the lesser of two evils' is normally consequentialist. Even economic discussions about the importance of increasing GDP is a version of consequentialist thinking, since wealth is often taken to be an indicator of people's ability to satisfy their preferences.

While consequentialism is normally portrayed as standing diametrically opposed to deontology, the two approaches are not always as separate as it would initially appear. For example, it is possible for a consequentialist to seek to maximize the number of rights that are protected. Rights are by their nature the province of deontology, but the consequentialist can use them as his or her measure of the desirable state of affairs being sought. It is also possible for a deontologist to have consequentialist constraints on their approach.[6] For example, a deontologist might act on the principle of protecting endangered species up to a point at which the costs become too high. Alternatively, they might seek to maximize happiness under the side-constraint of not sacrificing innocents in the process, or not restricting the fundamental liberties of others.

In sum, consequentialism has several strengths to it. It does not normally require strict adherence to principles that each require justification, instead asking simply that we act to create desirable states of affairs in the world. Furthermore, the measure of whether an action is right is, in principle, straightforward; did the act promote those desirable consequences or did it reduce them? Finally, consequentialism also fits with the widely held intuition that morality is about creating results that make the world into a better place.

Deontology and Consequentialism in Science and Technology Policy

In decisions about science and technology policy, deontological and consequentialist arguments are often both equally in evidence. Philosophical attention is often required to discern the different lines of argument. In the case of biotechnology, some people think genetic modification is wrong on deontological grounds, others think the potentially beneficial consequence of increased food security justifies any discomfort about the means. On the question of what to do about climate change, some seek to compare the consequences of doing nothing about climate change to the consequences of imposing a tax on carbon. At the prospect of geoengineering the climate, some say "no" on both consequentialist and deontological grounds; the former suggesting attempts to deliberately manipulate the climate will result in great harm, the latter suggesting it is fundamentally wrong to intentionally alter earth's biogeochemical system on such a grand scale. In the case of nanomaterials, the potential harms of nanoparticles for the body and the environment have caused some to call for precautionary regulatory action to be taken on consequentialist grounds. Others oppose on deontological grounds any such fundamental manipulations of matter.

It is not usually possible to pick a winning argument based

simply on whether it is deontological or consequentialist. Both have their power in different situations. In general, however, since technological development itself is normally seen as inherently neutral, as long as the costs and benefits are favorable, then consequentialist arguments often carry the day when it comes to determining how to proceed with new technologies. Not often are deontological arguments deemed decisive against the development of a new technology. A good illustration of an exception to this rule is the deontological argument against the cloning of human beings. In this case, a line in the sand has apparently been drawn based on the idea of the moral significance of the human genome. Deontological thinking, some say, prohibits human cloning.

The most important reason to become familiar with the difference between deontology and consequentialism is that identifying which type of moral argument is being used can be important for knowing how to evaluate and respond to it. The validity of consequentialist arguments for a certain action will usually hinge on the accuracy of the projected consequences of that course of action. This projection of consequences should normally be justified on empirical grounds and science often has in important role to play in supplying this empirical information. A policy maker might simply be empirically wrong that a certain course of action will result in beneficial (or harmful) effects. An illustration of this is how the presumption in favor of ethanol production on US cropland as a carbon mitigation strategy shifted when more became known about the petroleum intensity of producing biofuels. The consequentialist argument in favor of biofuels (i.e., reduced carbon footprint) turned out on closer inspection to be empirically mistaken. The consequences were not as projected, so the argument had to be qualified.[7]

In contrast, opposition to a particular course of action based on deontological principle is not always easy to dislodge empirically. As mentioned above, deontological principles tend *not* to depend on circumstances so, unless you can show that the principle should not apply to the situation in question, it will take a different kind of argument, getting into deeply held values, to persuade a person to change their position. For example, while science provides some useful information about when a fetus becomes viable outside of the mother's womb, the bigger question of whether a fetus is 'a person' resists an empirical answer.

c) Virtue Ethics

The third major approach to morality in the Western tradition is very different from the other two, and has tended to play only a marginal role in debates about science and technology. Rather than placing the focus on action and the reasons why certain acts are right or wrong, the virtue ethicist looks instead at the person who is acting and asks whether they are demonstrating appropriate moral character. For example, instead of putting the emphasis on whether a person should tell a lie, the focus is on whether their character is honest. Instead of looking at whether a person aided someone in distress, the virtue theorist asks what it would take to demonstrate courage. Naturally, a person becomes courageous (and certainly gets such a reputation) through doing courageous acts. However, in the virtue approach, the measure of moral goodness is not action itself but the settled character of the individual.

Virtue ethics has its origin in Greek thought and particularly the work of Aristotle. The reason Aristotle looked for the roots of morality in character rather than in action was that he did not think it made any sense to talk about an action begin good unless one had an idea of what sort of acts humans were supposed to be doing. For Aristotle, true knowledge of anything required a firm grip on the full nature of a thing. To do ethics, one needed to know something about the human purpose and the human function. With an accurate account of the overall human function and purpose in hand, it should be possible to determine those character traits that indicated someone was performing the human function well. Aristotle came up with a list of virtues including courage, friendliness, intelligence, generosity, and honesty, that he thought necessary for humans to fulfill their potential. His approach involves more of a long-term vision of persons learning how to refine their character than it does a moment-by-moment determination of whether a particular action is right or wrong.

Virtue ethics has gained popularity in the last couple of decades in Western philosophy. Initially, this might seem a little surprising. One of the strengths of both deontological and consequentialist ethics is that they provide mechanisms for definitively settling the question of how to act. Virtue ethics, by contrast, appears only to offer more general advice about good character and leaves out the specifics of particular actions. Its advocates counter that this is one of the system's strengths. Virtue ethics refuses to embrace the idea that moral decisions can be determined in the abstract. It does not divorce the morality of behavior from details about the individual doing it and the circumstances in which the individual finds him or herself. Context is always important. For example, the rightness or wrongness of attempting to tackle a robber running down the street might depend on whether you are an NFL linebacker or an elderly person with fragile bones.

A second major advantage of virtue theory is that it acknowledges that doing the right thing is not a purely mathematical operation nor is it a matter of simple obedience to moral law. It requires a number of admirable traits in a person to discern virtuous behavior. These traits include perceptiveness, judgment, sensitivity, and determination. Moral behavior is not simply the application of an algorithm to a situation. It requires what virtue theorists call "practical wisdom." Aristotle wanted to encourage the capacity to think about things in a certain way, rather than just to learn the contents of a rulebook. In the real world, rulebooks sometimes have only limited use. The idea of developing certain desirable character traits seems like a good way to teach morality.

The language used by virtue theorists is usually fairly recognizable. When admirable character traits such as courage, honesty, and generosity are championed, virtue ethics is in play. When a person giving moral advice stands back from saying specifically what should be done, but rather suggests in more general terms what might be expected of a person, virtue ethics is again at work. Another recognizable feature of the approach is the frequent use of role models to illustrate good character. When the memory of Mother Teresa's selfless charity is invoked, or the courage of Martin Luther King, or the vision of John F. Kennedy, it is clear that certain traits are being invoked, rather than a call to repeat specific acts. This use of role models is also a hint that there is more room for culturally different expectations in virtue ethics than in deontology or consequentialism.

It should be noted that virtue ethics cannot always be cleanly separated from deontology and consequentialism. A person who is meticulous about following the rules set down by a deontological approach could easily gain the reputation of being virtuous. Kant himself, the prototypical deontologist, wrote a book titled the *Doctrine of Virtue* that illustrates this possibility. Similarly, consequentialists have always been quick to suggest that virtues such as courage are virtues precisely because they have a tendency to lead to the best consequences overall.

Since virtue theory points not toward stringent rules for action but more toward the need to cultivate certain personal qualities, it may not always be as helpful a guide in deliberations over science and technology policy as consequentialism or deontology. Virtue theory tends not to give specific advice about whether or not to pursue a particular research agenda or technology. At the very least, however, those engaged both in the practice of science and in deliberations over science and technology policy should always be sure to demonstrate virtues such as openness, fairness, prudence, and intelligence in their work.

ETHICS AND THE ENVIRONMENT

Each of the approaches to ethics discussed so far have traditionally been concerned more or less exclusively with the relationship of one human being to another (or collections of others). For the first two millennia of Western philosophy, it was not unreasonably thought that these human-centered concerns exhausted the range of appropriate topic areas for ethics. In the last fifty years, this assumption has faced a concerted challenge. In 1949, one of the initiators of this challenge, Aldo Leopold, called for an "extension of ethics" to include the land, the soils, and all the biota.[8]

A new brand of environmental ethics—inspired initially by Leopold—has argued that adequate environmental protection will only emerge when the moral significance of natural entities in their own right is acknowledged. If a polar bear, a giant sequoia, or even an ecosystem as a whole has its own right to exist, then there is a deontological reason to protect them without any reference to human interests at all. Because this type of argument does not center itself on the interests of humans, it is known as a "nonanthropocentric" position. Both nonanthropocentric deontological arguments and nonanthropocentric consequentialist positions are possible. "Marine biodiversity has inherent value!" is an example of the former. "Tigers need undisturbed forest ecosystems to flourish!" is an example of the latter. The former invokes the rights and the natural values inherent in marine ecosystems calling for a principle of respect for biodiversity. The latter points towards the interests (i.e., happiness!) of tigers in having adequate habitat.

In recent years, environmental virtue theorists have joined the ranks of environmental ethicists. One environmental virtue position takes traits that have been determined virtuous for humans in their relationships with each other and shows how those same traits can create desirable interactions with the environment. For example, sensitivity, respect, aesthetic awareness, and the capacity to love are all traits that can be directed toward the natural world, with a good deal of environmental protection resulting. Another option for environmental virtue ethicists is to champion virtues that offer benefits primarily to their human possessors but also to the environment. The virtue of moderation, for example, prevents the possessor from personal excess thus saving their own character. But at the same time, moderation will also prevent a person from consuming all the environmental resources around them. In these and other ways, modern environmental virtue theorists are starting to show how desirable character traits and environmental concern are integrally linked.

PLURALISM, MONISM, APPLIED AND PRACTICAL ETHICS

As can be seen from the brief discussion above, there is no shortage of ways that ethics has bearing on social and environmental concerns related to science and technology. Utilitarian thinking appears in some discussions, deontological thinking in others. Virtue orientations can hover in the background even when deontology or utilitarianism is in the fore. This is the nature of the ethical beast. As Aristotle warned nearly two thousand four hundred years ago, one should only expect as much precision as the nature of a subject allows. Philosophical ethics is not a precise science and so considerable complexity remains in moral decision-making. Solving an ethical problem is rarely like solving a differential equation. It is not often that a moral problem encountered in the science debates can be adequately addressed from entirely within a single ethical framework.

As a result of this complexity, there has been considerable debate in environmental ethics about whether the goal should be to choose a single moral orientation and stick with it (monism) or whether one should be free to choose different moral frameworks for different parts of a moral problem (pluralism). J. Baird Callicott has argued for moral monism, worrying that the failure to settle on a single framework can lead to, at best, uncertainty, and at worst, inconsistency.[9] The cost of what Callicott calls "metaphysical musical chairs" is a public who can settle no environmental argument and individuals who are inconsistent in their positions. For example, an individual might be utilitarian about animal pain (wanting to minimize it) and deontological about the health of an ecosystem (wanting on principle to keep a native ecosystem healthy). This individual would be left helplessly confused about whether the eradication of nonnative goats from Olympic National Park was acceptable or not.

Other environmental philosophers such as Christopher Stone and Andrew Light have argued that environmental problems must be approached pluralistically.[10] One might need to be a virtue ethicist in one arena, a utilitarian in another, and a deontologist in a third.

Moral pluralism is more pragmatic, making the overall consistency of a worldview a secondary consideration to getting a problem solved. Problems in the environmental arena are multilayered, pluralists contend, making it unreasonable to request that a person choose a single philosophical orientation to apply to every possible scenario.

This pragmatist thinking has led to a second important distinction within environmental ethics. Bryan Norton, a philosopher now teaching in a school of public policy, has pointed out a difference between "applied" and "practical" ethics.[11] Applied ethicists conceive of ethical problem solving as a matter of carefully assaying a problem from afar before returning to the philosophy department to select an ethical theory "off the shelf" to solve the problem. The marine professor described in the anecdote at the start of this chapter was operating with a vision of applied philosophy in mind. The problem with such a vision is that it is rare for something off the shelf from a philosophy department to have the requisite detail and nuance to solve a particular real-world problem. A practical problem is only occasionally a suitable test case for a set of abstract principles.[12]

Practical ethicists know that environmental dilemmas may be construed differently by people coming at them from different value orientations. Wolves in the rural west may be in part a livelihood issue for a rancher but an aesthetics issue for a photographer. You are unlikely to find a single theory on the shelf that can resolve the question of how to manage the wolves for these two parties. A practical approach to ethics starts from a pluralistic framework and tries to hone in on a conception of a common future that could work for both parties. Practical ethics employs theories only to the extent that they can contribute toward working out the common ground. The discussion that ensues has education and the articulation of common values as its goal.[13] Practical environmental philosophers are more likely to look for ways of forging an acceptable compromise rather than solving the question of who "wins" a particular moral debate based on the power of their argument.

ETHICS AND THE SCIENCE DEBATES

This short overview by no means exhausts the tools that philosophical ethics can bring to the science debates. The three main traditions in Western ethics are only the tip of an iceberg with many overlapping layers, diffuse edges, and ever-changing accretions. Chunks of this iceberg sometimes break off and become part of other fields such as environmental politics or sociology. An important example is environmental justice (EJ), a movement concerned about the disproportionate impact of environmental benefits and harms on marginalized or disempowered populations. Since it is firmly rooted in deontological discussions of rights and justice, EJ has clear roots in philosophical ethics even though few of its advocates reside in academic philosophy.

The advantages of being able to identify the basic orientations in philosophical ethics are numerous. They go far beyond simply gaining more confidence in deliberating in an interdisciplinary environment. As discussed above, familiarity with these traditions can provide clues about how to engage with those who offer a different opinion on a science and policy direction. Is this just a disagreement about facts, or are there deeply held principles contending with each other? Do we need to wait for more data, or is this a case where the information will always be imperfect? Is there any room for compromise here, or has a line been drawn in the sand? The rhetorical benefits of knowing this field are numerous.

It would be a mistake to suggest, however, that the purpose of knowing about these ethical orientations is simply to become better at arguing a position. Familiarity with philosophical ethics can also play an important role in allowing significant moral values to be identified and important policy directions to be refined. In deliberations over agricultural biotechnology, nanotechnology, and climate ethics, having a grip on the fundamental ethical orientations involved makes public debate more informed, more focused, and more likely to result in research directions that are more consistent and lasting. The consequentialist/deontology dance appears in all of these debates, with virtue theory emerging out of the background in the form of questions about what sort of a people we want to be and what sort of planet we want to leave to our grandchildren. These ethical considerations remain deeply relevant to both the practice and the direction of science.

Knowing a little about philosophical ethics allows scientists to know the lay of the ethical landscape that their work invariably helps to shape. Moral theories are not tools to be applied by distant experts in philosophy departments to solve the large practical challenges facing humanity. Moral problems are almost always about finding common ground after considering varying costs and benefits, ensuring that those affected are treated justly, and showing desirable character traits when solving the difficult public challenges. Knowing what is at issue ethically is essential if scientists are going to avoid surrendering this territory to those who lack the first-hand knowledge of the earth systems that are at stake and the technologies that might be used to protect them. Students and practitioners in science and engineering will benefit from using the structure provided by philosophical ethics to help them identify the role they can play as citizens in securing the right kind of future for humanity and for the Earth itself.

4.

BRIDGING THE GAP

Global Justice in Health Research

Julian Culp and Nicole Hassoun

1. INTRODUCTION

There is a large gap between global health needs and current health research.[1] Most of the funds for health research are in developed countries and respond to their health priorities, while developing countries bear the largest disease burden.[2] More generally, there may be similar problems with the way scientific research is organized in other areas. Some ways of reducing gaps between scientific research and need may be better than others. This article argues that, together, ethical and scientific inquiry may help determine which ways are best.[3] It considers, for instance, a few ways ethical and scientific inquiry is useful in creating, evaluating, and implementing proposals to narrow the gap between health research and need. Further, it argues, there are many things scientists, in particular, can do to address this problem. Leaders from the scientific community can suggest policy changes to broaden research priorities and engage with research questions that are relevant for assessing new proposals to reduce the health research gap. They can share their expertise with scientists in developing countries. Although no way of addressing the global health research gap is perfect, ethical and, especially, scientific inquiry may be essential to making some significant progress. More generally, dialogue between ethicists and scientists may be helpful for addressing other problems with the way global scientific research is organized.

2. THE GLOBAL RESEARCH GAP AND THE CASE OF HEALTH RESEARCH IN PARTICULAR

The 1990 landmark report *Health Research—Essential Link to Equity in Development* revealed the failure of global health research and development (R&D) funds to target humanity's most severe health problems.[4] It revealed that, although developing countries bore 93 percent of the global disease burden (GDB),[5] only 5 percent of the global health R&D expenditures were directed to health problems in these countries.[6]

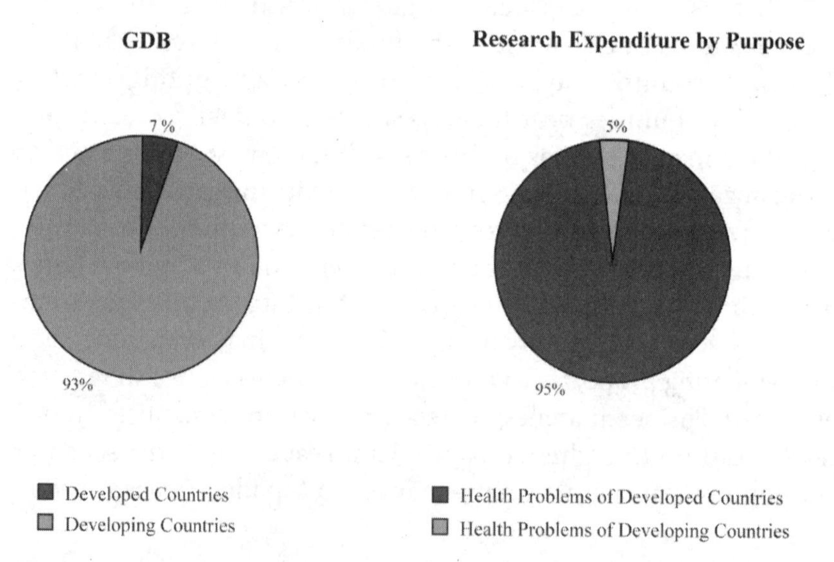

Figure 1: Contrast in GDB and allocation of health research funds.[7]

Founded in 1998, the Global Forum for Health Research coined the term "10/90 gap" to describe this imbalance in health R&D funding.[8] It found that in 1990 only 10 percent of global health R&D was directed to the health problems that caused 90 percent of the global disease burden.[9] Today, mainly through investments from the Bill and Melinda Gates Foundation, the imbalance has markedly decreased.[10] Nevertheless, the global health research gap remains very wide.[11] To illustrate the problem, consider the table below:[12]

Condition	GDB in Million DALYs	% of Total GDB	R&D Funding in US$ Millions	R&D Funding in US$ per DALY
All GDB	1,470	100	105,900	72
HIV/AIDS + TB + Malaria	167	11.4	1,400	8.4
CVD	148.19	9.9	9,402	63.45
Diabetes	16.19	1.1	1,653	102.07
HIV/AIDS	84.46	5.7	2,049	24.26
Malaria	46.49	3.1	288	6.2
TB	34.74	2.3	378	10.88

Figure 2

The Global Forum for Health Research highlights that in 2001 105.9 billion US dollars (105,900 million) were spent on global health R&D. Thereof, roughly 1 percent was spent on diabetes. Similarly, about 1 percent was spent on HIV/AIDS, tuberculosis, and malaria. These investments fail to track the most pressing global health problems. For while diabetes was responsible for 1.1 percent of the GDB, HIV/AIDS, tuberculosis, and malaria together made up 11.4 percent of the GDB. More recent studies find that tropical diseases and tuberculosis account for 12 percent of the GDB and cardiovascular diseases account for 11 percent. Between 1975 and 2004, however, only 21 new drugs were indicated for tropical diseases and tuberculosis, while 179 new drugs were developed for cardiovascular diseases.[13] Global health R&D is extremely sensitive to the health problems of developed countries and fails to address the health needs of those in developing countries.

In general, there might be a significant research gap between global R&D investments and global need for the fruits of scientific research. Developed countries account for more than 75 percent of gross domestic expenditure on scientific R&D, although they represent only about one fifth of the world population.[14] That is to say that developed countries spend roughly twelve times more per capita on R&D than developing countries. More than 60 percent of the world's researchers work in developed countries.[15] Since much global R&D is done by private, for-profit companies[16] and most of the money is in developed countries, it would not be surprising if most of the world's R&D addressed the priorities of these countries.[17] Furthermore, even when R&D is funded by public entities, it may not address the needs of developing countries. Again, using the example of health research, consider that in 2008 the US government spent only 4.3 percent of its total health R&D budget on neglected diseases that affect a large segment of the world's population.[18] Notably, the United States is by far the biggest public spender on R&D on neglected diseases. It contributes 67.2 percent of public and 42.6 percent of total global funding for R&D on neglected diseases.[19]

Consider the state of global agricultural research and how such research could influence rural poverty. The 2008 *World Development Report* suggested that some of the poorest, especially in African countries, have received little benefit from recent advances in biotechnology.[20] It suggests that most research in biotechnology is done by the private sector and is not primarily directed toward the interests of the poor.[21] Yet, most poor people live in rural areas and depend for their livelihoods on agriculture. Many poor people could benefit, for instance, from drought resistant crops.[22] Agriculture is also very important for poor countries' economic development.[23] So, there may be a global agricultural research gap as well as a global health research gap.

3. THE NEED FOR ETHICAL AND SCIENTIFIC DIALOGUE

Together ethical and scientific inquiry may help address pressing global problems like the global health-research gap and any other existing research gaps. Even on the assumption that such gaps are morally problematic, it is not clear what to do about them.[24] Some ways of reducing these gaps may be better than others. Together, scientific and ethical inquiry may help researchers create, evaluate, and implement proposals that are not only feasible but also ethically sound.

Consider how scientific and ethical inquiry may help researchers evaluate policy proposals for reducing global research gaps. Scientific inquiry may help determine the origin, development, and effects of such gaps. It may isolate the causal factors that create and sustain these gaps and their direct and indirect effects. Ethical inquiry may be useful in determining which ways of reducing global research gaps are better on ethical grounds. Ethicists may argue for a list of desiderata that good policies must fulfill. They may conclude, for instance, that good ways of narrowing this gap must be efficient and/or equitable. Ethicists can also consider how different desiderata should be taken into account in an overall assessment of a given policy. Even if, say, both efficiency and equity matter, one may matter more than—or trump—the other. Together ethical and scientific inquiry may allow researchers to determine which particular proposals will score highest on all the relevant ethical desiderata.

In the case of the global health research gap, for instance, researchers might begin from the desideratum that health research should ameliorate the GDB, especially for the poorest. Scientific inquiry may clarify the effects of alternative strategies with respect to this objective, and ethical scrutiny may allow researchers to decide between them. Scientists might evaluate the incentive structures that lead to the global health research gap in the first place. They may discover that one proposal is likely to alleviate the total disease burden more efficiently. Another may be better for the poorest of the poor. This is where ethical deliberation may be necessary to properly balance these different considerations. Those who favor the former proposal may want to implement the policy that does best in reducing the total disease burden. Those who prefer the latter may opt for the strategy that contributes most to reducing the disease burden on the poorest. Ethical theories can provide reasons for or against either of the two proposals. Together scientific and ethical inquiry can initiate and facilitate debates that may ensure good policy decisions. In the next section we draw on actual policy proposals to further clarify how dialogue between ethicists and scientists can be useful in creating and implementing as well as evaluating these proposals. Moreover, we highlight ways in which scientists may be particularly well placed to articulate, evaluate, and implement policies for reducing the health research gap.

4. POTENTIAL SOLUTIONS AND THE ROLE OF ETHICAL AND, ESPECIALLY, SCIENTIFIC INQUIRY

This section discusses the many important roles that ethicists and, particularly, scientists play in creating, evaluating, and implementing different strategies for reducing the health research gap. To make this case, it considers just a few potential solutions to the health research gap and explains how good proposals have been—and might be—created, evaluated, and implemented.

Consider, first, Aidan Hollis and Thomas Pogge's proposal to create a new, voluntary, patent scheme.[25] Their central idea is to reward innovators for their contributions to reducing the GDB. Scientists would register a new innovation under this alternative scheme instead of registering it under the current patent system. Rather than recouping the R&D investments by profiting from a monopoly for a limited period of time—as companies do under the current intellectual property rights system—inventors would receive financial benefits from a fund in proportion to the positive impact of the product on the GDB. Producers participating in this second patent scheme would not have an incentive to sell the end products at high monopoly prices. Rather, they would have an incentive to sell them close to, or even below, marginal costs to increase innovations' impact on the GDB. The financial resources necessary to pay companies for their innovations' impact would have to be provided by public and private bodies. Initially, Pogge estimated that the plan "might cost some US$45–90 billion annually on a global scale," which would amount to 0.1 to 0.2 percent of the global product.[26] Later, Hollis and Pogge estimated that it would cost US$6 billion.[27]

The main appeal of this alternative patent scheme is that it may generate incentives to concentrate research efforts on those diseases where the costs for carrying out R&D are lowest in proportion to their expected impact on the GDB. On Hollis and Pogge's proposal, companies would decide which investments to make in trying to alleviate the GDB. So those implementing their system would not need to decide how much a given intervention is worth ahead of time.

Hollis and Pogge's proposal has some potential drawbacks as well. First, it may be extremely difficult to determine how much a given invention has contributed to reducing the GBD. The invention of a new biomedical product may only reduce the GDB in conjunction with other measures that a government or an international organization carries out. Some vaccines, for instance, require refrigeration and this may require upgrading the health infrastructure of some developing countries. Second, the history of unfulfilled financial promises to help the world's poorest raises serious doubts about whether it is feasible to secure adequate funding for the proposal.[28]

Hollis and Pogge's proposal illustrates how, together, ethical and scientific inquiry can be useful in creating, evaluating, and implementing strategies to narrow the health-research gap. It was only with interdisciplinary dialogue that the proposal was created and developed.[29] Together ethical and scientific inquiry may allow researchers to evaluate this proposed solution to the research gap. Scientific inquiry can help determine how seriously to take any of the potential advantages or disadvantages set out above. Social-scientific inquiry, for example, might help researchers determine how politically feasible it would be to secure the funding that Hollis and Pogge say they need, or how likely it is that the funding will generate sufficient incentive for scientists to create new innovations. Ethical inquiry may help researchers decide whether the focus on global health impact is appropriate. It might be better to revise the proposal so that it gives companies larger rewards for alleviating the health burden among the poorest, for instance. Furthermore, it is clear how scientific inquiry will be important in implementing Hollis and Pogge's proposal. Scientific inquiry is necessary to measure new innovations' impact on the GDB. This requires empirically estimating the global disease burden's size as well as any newly developed medicine's impact. This is no small task given that different conditions exist in each country, e.g., some have much better health infrastructure than others. Furthermore, measuring the GDB requires a health metric. Researchers might, for instance, use DALYs or Quality Adjusted Life Years (QALYs). Each health metric includes a specific valuation of a lost life year. Usually, life years in the youth of a person are weighed more heavily than in seniority. This, however, is an ethical judgment that requires defense. Ethical inquiry may suggest various considerations in defense of one particular view about the proper weight of the loss of a given life year. So such inquiry may support some health metrics over others.

Besides Hollis and Pogge's proposal, there are many other initiatives that demonstrate how scientific inquiry, in particular, can contribute significantly to addressing the health research gap. Scientists, for example, are particularly well placed to promote policy changes to broaden research priorities. Alternately, their research results may provide relevant information for assessing different policy proposals. Finally, they can act as agents of social change by helping to implement policies that reduce the health research gap.

The University of California Berkeley's Socially Responsible Licensing Program provides a good example of how scientists can initiate changes in the way research is structured.[30] In 2002, Eva Harris at the School of Public Health, along with her colleagues in the department of Electrical Engineering and Computer Science, managed to successfully implement the Socially Responsible Licensing Program. This program changed the way Berkeley measured the success of its licensing office. Formerly, the revenue from the sale of licenses and the impact of the sale on local economic development were the relevant indicators. Therefore, the office was not rewarded for selling licenses at a low price for the production of pharmaceuticals to benefit people in developing countries. In the new evaluation metric, Berkeley considers the impact of licenses on the affordability and accessibility of medicines in developing countries. Thus social as well as financial impact indicators now constitute the metric for assessing the licensing office's performance. This so-called double bottom line concept for valuing the office's work may increase research on neglected diseases. It can reward the office for granting licenses that do not increase the university's revenues. The office received credit, for instance, for partnering with the Sustainable Sciences Institute to support the development of a handheld micro-electro-mechanical systems-based diagnostic for the diagnosis of dengue fever. The Sustainable Sciences Institute was then able to distribute the diagnostic for free, since it did not have to recoup the cost of buying the license. The program has also had other good effects. It enabled Berkeley to collaborate with several not-for-profit organizations, including the Bill and Melinda Gates Foundation, to harness the university's expertise in health research to help poor people.[31] Today, many other universities are in the process of changing their license schemes.[32]

Scientific research also plays an important role in evaluating strategies to address the health research gap. Consider, for instance, the World Health Assembly's "[g]lobal strategy and plan of action on public health, innovation, and intellectual property."[33] Among other things, it demands promoting developing countries' R&D capacity on neglected diseases via technology transfer from developed countries.[34] This strategy is based on the WHO's report *Public Health—Innovation and Intellectual Property Rights*.[35] The report states, "Many [developing] countries, particularly the more scientifically advanced, have positive advantages as low-cost producers of high quality product."[36] This assertion, in turn, is supported by a study that presents the potential savings from technology transfers to—and local production in—developing countries.[37] The expected impact of technology transfers may, however, be critically scrutinized and compared with other approaches to addressing the gap in health research. Some authors carrying out scientific research for the

World Bank question whether supporting local production capacities could reduce the gap in R&D. They argue that "producing medicines domestically makes little economic sense. If many countries begin local production, the result may be less access to medicines, since economies of scale may be lost if there are production facilities in many countries."[39] This controversy shows that scientific research fulfills the very important task of properly evaluating alternative strategies. Even if this inquiry yields disagreement, there is often no better way to assess the desirability of a given proposal than to rely on scientific expertise.

Finally, there are many ways in which scientists may be particularly well placed to help implement proposals. Scientists with expertise and influence in industry might, for instance, participate in or initiate corporate programs to train people in developing countries. Scientists at the pharmaceutical company GlaxoSmithKline (GSK) can, for example, help promote technology transfer by training scientists in Shanghai. These programs might be expanded so that scientists from many countries can share best practices by spending some time at each other's laboratories. After all, GSK already partners with the UK India Education and Research Initiative (UKIERI) that was launched in 2005. The initiative aims to enhance academic and scientific exchange and, among many other things, enables Indian students to spend one year in GSK's R&D divisions to learn from GSK's scientists.[39] In these ways scientists can help advance R&D activities in developing countries.

Scientists can also raise awareness of the global health research gap and advance good proposals for addressing it by actively engaging in public discourse. Consider the work of *Medécins Sans Frontières*, one of the world's largest private humanitarian relief organizations, and the Drugs for Neglected Diseases (DND) Working Group. Together these organizations significantly increased public awareness of the health research gap by publishing their influential report "Fatal Imbalance."[40] Scientists played an important role as members of the DND Working Group and the editorial advisory board of the report.[41] The work of this group also resulted in the foundation of the Drugs for Neglected Diseases initiative that, apart from developing new drugs, advocates for greater political attention to neglected diseases.[42]

5. CONCLUSION

This paper suggested that there may be some problems with the way scientific research is organized and that dialogue between ethicists and scientists may be helpful in addressing these problems. It considered, for instance, how dialogue between scientists and ethicists has been useful in creating, evaluating, and implementing proposals to narrow the gap between health research and need. Further, it argued, there are many things scientists, in particular, can do to address this problem. Thought leaders from the scientific community can suggest policy changes to broaden research priorities. Scientists can also engage with research questions that are relevant for assessing new proposals to reduce the health research gap. Alternately, they can share their expertise with scientists in developing countries. Although no way of addressing problems with the way scientific research is organized is perfect, ethical and scientific inquiry may be essential to making some significant progress. More generally, dialogue between ethicists and scientists may be helpful for addressing many pressing global problems.

Section 2

POLICY AND THE SCIENCE DEBATES

5.

INTELLECTUAL LIBERTY AND THE PUBLIC REGULATION OF SCIENTIFIC RESEARCH

Clark Wolf

I. FORBIDDEN KNOWLEDGE AND POLITICAL SCIENCE

Calls to regulate or restrict scientific research are often a matter of politics, and public desire to regulate science may have its source in several different underlying interests: on one side, people may be motivated by an interest to control risks, prevent harms, or limit access to powerful or dangerous technologies. These interests are easy to understand, and often provide entirely appropriate and creditable grounds for regulation. In a darker vein, people may be motivated by more general mistrust of science, or by moral or religious disapproval of some kinds of research. While these motives may be easy to understand, clearly they should be resisted. But if researchers hope to avoid inappropriate regulations, we need to be prepared to explain our research to the public. And when research is funded by tax dollars, it is especially important that scientists should be able to justify its benefit to taxpayers and legislators.

Regulation of science can either promote or inhibit research. Either way, things can go well or badly. When Galileo's great work on celestial mechanics was suppressed because his results were inconsistent with the teachings of the church, we recognize this as censorship. And we recognize as similarly bogus Trophim Lysenko's receipt of the Stalin Prize for his ideologically informed and bogus "research" on agronomy and crop genetics. Contemporary US policy also promotes and inhibits research in select areas: special incentives are available for scientists whose work promotes "clean energy" (itself a questionable term that gets its meaning from nonscientific legislative fiat), and there are disincentives for research on human embryonic stem cell lines. Here as in other cases, regulation limits research and inquiry. Since freedom of inquiry and freedom of intellect are among the most precious liberties we possess, it is relevant to try to specify the circumstances (if any) in which these liberties can appropriately be constrained and identify very clearly the interests (if any) that justify such constraint. Those who are engaged in scientific research often regard public scrutiny and regulation as an unwarrantable intrusion and as the inappropriate incursion of political interests where they do not belong. A theory of regulation should explain when this attitude is justified and when it is not.

In science, we can have some confidence that the truth will eventually rise to the top, even when science is regulated and inquiry is curtailed. Today we celebrate Galileo's brilliance and vision and condemn Lysenko as a pseudoscientific fraud. But our confidence about these judgments should not engender a complacent attitude toward regulation. Some contemporary examples of the political regulation of science are quite as contentious as those of Galileo and Lysenko: In the United States, politics have entered the research process in a variety of different ways, specifying which kinds of research may be done with public funds and at public universities (example: stem cell research), how research is communicated (example: global climate change research), and how new technological developments are made available (example: RU-486, the "abortion" drug). This paper begins with questions that should be taken seriously by anyone engaged in scientific research:

What business do nonscientists have to regulate science?

What legitimate interest (if any) do legislators and the public have to restrict and regulate the professional activities of scientists engaged in research?

These questions are especially pressing, since laypeople and lawmakers typically do not understand the science they hope to regulate. Consequently, their regulations may often be badly framed and poorly conceived. This lack of understanding, however, is not just the unfortunate result of ignorance or inadequate science education. Most cutting-edge research is complicated and specialized: even scientists working in the same field often have difficulty understanding the work of researchers in an adjacent laboratory. It is inevitable that regulations governing scientific research will be framed and approved by people who do not fully understand the research in question. Such regulations do limit the liberty of scientists and their ability to pursue their goals, and they have a decisive influence on which research will be undertaken and how it will be pursued.

In some cases, such regulation constitutes "political science" in the worst possible sense: if the interests that guide regulations are inappropriate or ideological, the result can be science that is inappropriately influenced by political or other interests. Just as "scientists" working for the US tobacco companies produced "research" to undermine evidence that smoking causes cancer, there have been more recent events in which political interests inappropriately influenced climate research.

Many observers regarded the Bush administration's efforts to regulate human embryonic stem cell research as a clear case of inappropriate incursion of political ideology into the research process. Barack Obama later issued an executive order (Executive Order No. 13,505, Fed. Reg. 10,667, March 9, 2009) in which he attempted to remove the barriers that prevented federal funding for this research, but his order was challenged in court and subsequently ruled to be illegal. In a controversial ruling, Chief Justice Royce Lambeth, of the US District Court for the District of Columbia ruled in August 2010

An Editor in the White House

Handwritten revisions and comments by Philip A. Cooney, chief of staff for the White House Council on Environmental Quality, appear on two draft reports by the Climate Change Science Program and the Subcommittee on Global Change Research. Mr. Cooney's changes were incorporated into later versions of each document, shown below with revisions in bold.

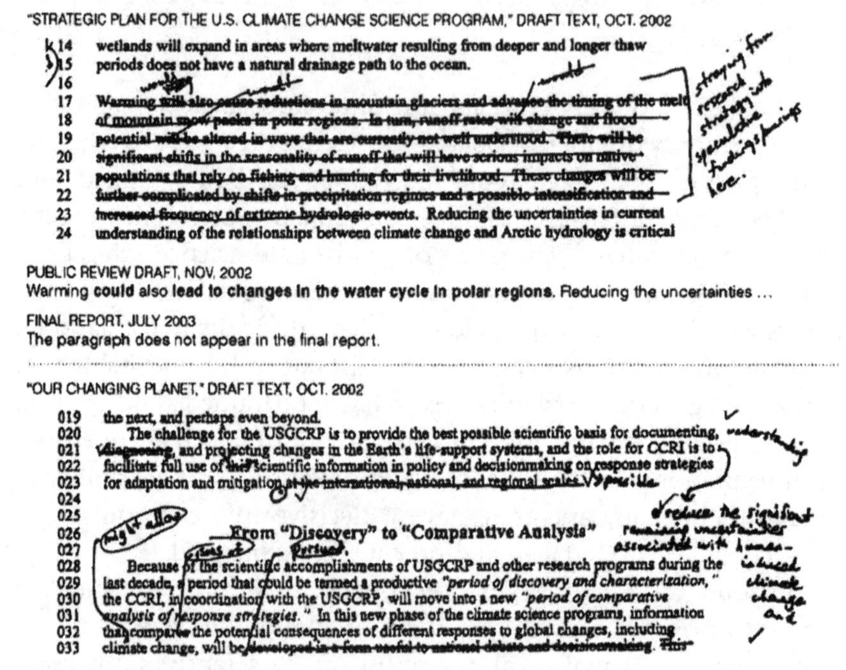

"STRATEGIC PLAN FOR THE U.S. CLIMATE CHANGE SCIENCE PROGRAM," DRAFT TEXT, OCT. 2002

14 wetlands will expand in areas where meltwater resulting from deeper and longer thaw
15 periods does not have a natural drainage path to the ocean.
16
17 Warming will also cause reductions in mountain glaciers and advance the timing of the melt
18 of mountain snow packs in polar regions. In turn, runoff rates will change and flood
19 potential will be altered in ways that are currently not well understood. There will be
20 significant shifts in the seasonality of runoff that will have serious impacts on native
21 populations that rely on fishing and hunting for their livelihood. These changes will be
22 further complicated by shifts in precipitation regimes and a possible intensification and
23 increased frequency of extreme hydrologic events. Reducing the uncertainties in current
24 understanding of the relationships between climate change and Arctic hydrology is critical

"OUR CHANGING PLANET," DRAFT TEXT, OCT. 2002

019 the next, and perhaps even beyond.
020 The challenge for the USGCRP is to provide the best possible scientific basis for documenting,
021 diagnosing, and projecting changes in the Earth's life-support systems, and the role for CCRI is to
022 facilitate full use of this scientific information in policy and decisionmaking on response strategies
023 for adaptation and mitigation at the international, national, and regional scales.
024
025
026 From "Discovery" to "Comparative Analysis"
027
028 Because of the scientific accomplishments of USGCRP and other research programs during the
029 last decade, a period that could be termed a productive *"period of discovery and characterization,"*
030 the CCRI, in coordination with the USGCRP, will move into a new *"period of comparative*
031 *analysis of response strategies."* In this new phase of the climate science programs, information
032 that compares the potential consequences of different responses to global changes, including
033 climate change, will be developed in a form useful to national debate and decisionmaking. This

PUBLIC REVIEW DRAFT, NOV. 2002
Warming **could** also **lead to changes in the water cycle in polar regions.** Reducing the uncertainties ...

FINAL REPORT, JULY 2003
The paragraph does not appear in the final report.

FINAL REPORT, 2003
The challenge for the USGCRP is to provide the best possible scientific basis for documenting, **understanding**, and projecting changes in the Earth's life-support systems, and the role for CCRI is to **reduce the significant remaining uncertainties associated with human-induced climate change and** facilitate full use of scientific information in policy and decisionmaking on **possible** response strategies for adaptation and mitigation.

Figure 1[1]

that Obama's executive order violates the Dickey-Wicker Amendment of the appropriations bill for Health and Human Services (US District Court for the District of Columbia, Civ. No. 1:09-cv-1575 [RCL], August 23, 2010). This amendment prohibits the federal use of funds for "(1) the creation of a human embryo or embryos for research purposes; or (2) research in which a human embryo or embryos are destroyed, discarded, or knowingly subjected to risk of injury or death greater than that allowed for research on fetuses *in utero*" under other applicable federal regulations. While some hailed this decision as an appropriate interpretation of the law and as enforcing needed protections for fetal life, others regarded it as an inappropriate judgment in which Judge Lambeth allowed his private moral convictions to influence his legal judgment.

While we will consider the case of stem cell research later, we must recognize in general that regulatory restrictions do limit and frame research, and as such they limit the liberty of scientists. Such limitations can undermine the ability to pursue responsible research. Because of this, it is crucial to consider and evaluate the public reasons for such regulations.

II. THE CASE FOR SCIENTIFIC LIBERTY

It is easy to sketch the case in favor of the liberty to pursue scientific research without undue impediments: In part, the freedom to pursue scientific research is simply one aspect of the more general liberty we enjoy, or should enjoy as participants in a free society. But beyond this, freedoms of intellect and thought are special freedoms, regarded as fundamental in US constitutional law, and by many political theorists.[2] Still, while freedom in the practice of scientific research is closely related to intellectual liberty, it may be somewhat broader in scope: scientific practice is not typically *expressive* activity, and the liberty to pursue empirical research is distinct from, though perhaps no less important than, the liberty to believe what one's intellect and conscience may dictate. These principles of liberty of action, intellect, and conscience are very general considerations, and they apply not only to scientific research but also to all our actions and pursuits.

Beyond these very general considerations, there are special reasons why we might hope to give special protections to scientific enquiry: scientific advance and innovation are important public goods, and we have a strong public interest to promote them.[3] In the modern world, scientific advancements and the infrastructure necessary to generate them are important sources of culture, resources, and social wealth. To produce these benefits, scientists need resources, infrastructure, support, and liberty. The wealth and welfare of nations that are unable or unwilling to provide these necessities will be very seriously at risk as information and technology become increasingly important to global economy and culture.

But the notion that scientific knowledge is a public good is sometimes called into question. In the modern world, the most economically significant advances in scientific knowledge are usually private intellectual property. Privately owned and controlled knowledge might be thought to benefit its owners, not the general public. In response to this concern, it's worth noting that the *aim* of intellectual property law is itself to promote the public good: Intellectual property protections, in fact, are among the ways in which scientific knowledge is *regulated,* since regulations can be designed either to promote what is regarded as desirable or prevent what is regarded as dangerous or undesirable. But intellectual property rights are (supposed to be) designed to promote the public good by providing an incentive for innovation and research. This is specified in Article 1 Section 8 of the US Constitution, which grants Congress the power "to promote the progress of science and useful arts, by securing for limited times to authors and inventors the exclusive right to their respective writings and discoveries." The purpose of intellectual property (IP) law is to "promote progress of science and useful arts" by providing an incentive for those who pursue them. This underlying aim provides a standard that can be used to evaluate existing laws: where IP laws stifle or constrain innovation, they fail to serve the purpose for which they were enacted. But the Constitution only protects intellectual property rights that "promote the useful arts," and the constitutional reasons for patent legislation are framed to promote the public good. It is noteworthy that the Constitution *permits* the creation of patents but does not require it. The framers evidently viewed patent law, and intellectual property law more generally, to be justified by the public benefits they provide and not by any view that inventors and researchers have an independent nonlegal right to control the intellectual goods they produce. Where private intellectual property rights don't serve this objective, they raise both constitutional and legal concerns. As I will argue, this provides very good reason why scientists who use the US patent laws should feel (and fulfill) an obligation to ensure that their work really does have net public benefits. Scientists should also be prepared to explain these benefits to the public that pays a substantial portion of the research costs that produced, and which is supposed to benefit from, the institutions that create and enforce inventors' rights.

III. POLITICAL AUTHORITY, REGULATION, AND CONSTRAINT

Given the general right to liberty and the instrumental value of scientific advance, what considerations justify regulation? In US law it is the constraint of liberty, not its exercise, that requires justification. That is, in the absence of a sufficient and compelling reason in favor of constraint or regulation, people are understood to have a right to do as they please. This thought is sometimes called the "presumption in favor of liberty."[4] Under this presumption, if Alph wishes to engage in some behavior but Beth wishes either to prevent him from doing so, or to regulate him as he does so, the burden lies on Beth to show that there is a good and sufficient reason that justifies the limitation of Alph's liberty.

Should we accept this presumption in favor of liberty? In one sense, this presumption simply embodies what it means to live in a free society. Liberty is an important good, and the right to liberty is a fundamental value. United States legislatures and courts have recognized such a presumption in a variety of different contexts: The Ninth Amendment of the US Constitution specifies, "The enumeration in the constitution of certain rights shall not be construed to deny or disparage others retained by the people." Discussion surrounding the enactment of this amendment shows that the framers regarded it as necessary to insure that later legislators and courts would not assume that people *don't* possess a right simply because it is not specified in the Constitution.[5] The Constitution ceded a limited set of rights to the government, which then was understood to have only those powers that were expressly given.

In another context, John Stuart Mill articulated a powerful defense of individual liberty and the limits of regulatory legislation.[6] Mill famously writes:

> The only purpose for which power can be rightfully exercised over any member of a civilized community, against his will, is to prevent harm to others. His own good, either physical or moral, is not a sufficient warrant. He cannot rightfully be compelled to do or forbear because it will be better for him to do so, because it will make him happier, because in the opinion of others, to do so would be wise, or even right. These are good reasons for remonstrating with him, or reasoning with him, or persuading him, or entreating him, but not for compelling him of visiting him with any evil in case he do otherwise. . . . Over himself, over his own body and mind, the individual is sovereign.

The principle that limitations on liberty are permissible only when they aim to prevent *harm to others* is usually called the *Harm Principle.* Mill intended this principle, for which he provides a detailed defense, to provide a complete theory about the rightful limits of the coercive power of the state. Mill's view is plausible and has been defended by many contemporary theorists.[7] Applied to the context of scientific investigation and the public regulation of science, this view would imply that the *only* reason that can justify the regulation of scientific activity is the imposition of harm or risk of harm on others. Thus, on this view restrictions that specify the treatment of research subjects would qualify as legitimate, as would restrictions that regulate risky research that imposes risks on nonparticipants. If this were the sole justification for the restriction and regulation of scientific research, then many such regulations would be unjustified and unjustifiable.

IV. REASONS FOR REGULATION

It is useful to consider reasons that have actually been cited to justify the regulation of scientific research. This section will consider reasons based on the "Faustian Myth" (if myth it is) that science must be regulated to prevent scientists from "going too far," and the notion that research may be regulated when people regard the methods or intended result to be "immoral."

(i) Regulation to Prevent Scientists from "Going Too Far": Faust and Forbidden Fruit

One motive to regulate science may be found in public mistrust of scientists and public concerns about research or technology "going too far." This concern has found its expression in literary works: In Goethe's *Faust,* a researcher sells his soul to the devil in his effort to gain knowledge and mastery over nature. In Mary Shelly's *Frankenstein,* science gone awry literally produces a monster. Timothy Ferris has dubbed this notion that "science must be reined in lest it go to far" the "Faust Myth."[8] This myth, if a myth it is, stretches back to some of our oldest stories—to Adam and Eve and the forbidden tree of knowledge, and to the earlier stories that have been identified as progenitors of that story from *Genesis.* In one of its most appealing representations, this view appears in John Milton's epic poem *Paradise Lost.*[9] In an early section of Milton's poem, Adam and Eve are amazed to find themselves in the Garden of Eden. They ask for information about their circumstances and their origin, and the angel Raphael responds:

> I have received, to answer thy desire
> Of knowledge within bounds; beyond abstain
> To ask, nor let thine own inventions hope
> Things not revealed, which the invisible King,
> Onely Omniscient, hath suppresed in Night,
> To none communicable in Earth or Heaven:
> Enough is left besides to search and know.[10]

Raphael makes it clear that they will receive only as much knowledge as fits their limited capabilities, and that there are other realms of knowledge that have been reserved for God alone. But Raphael explains his reasons for this constraint:

Knowledge is as food, and needs no less
Her Temperance over Appetite, to know
In measure what the mind may well contain,
Oppresses else with Surfeit, and soon turns
Wisdom into Folly, as Nourishment to Wind.[11]

Explaining these lines, Wendell Berry writes: "Raphael is saying, with angelic circumlocution, that knowledge without wisdom, limitless knowledge, is not worth a fart; he is not a humorless archangel."[12]

The view that scientists "should not go too far" has contemporary defenders. Echoes of the view can be found in the works of some authors I admire very much, including Wendell Berry and Bill McKibben.[13] But as the underlying source of this concern, in many cases, these writers are animated by their sense that science has provided technical mastery that threatens harm. Thus McKibben is concerned that new technologies will harm people and the environment, and Berry is concerned to argue that economic "knowledge" of the workings of free markets is dangerous because, as he argues, it's essentially false and leads us to bad policy, not because it reflects an intemperate pursuit of essentially *forbidden* knowledge. Perhaps, in the face of motives to regulate scientific research to prevent scientists from "going too far" we should simply maintain our commitment to liberty and permit regulation only to prevent harm and risk of harm to others. It would clearly violate the presumption in favor of liberty and Mill's harm principle to constrain or regulate scientific research in order to prevent scientists from discovering knowledge that some may regard as hidden or forbidden.

(ii) Public Reasons and Regulation to Prevent "Immoral Science"

In other contexts, regulation seems to be based on a desire to prevent the pursuit of scientific projects that are regarded as "immoral." For example, efforts to regulate the development and distribution of RU-486, the so-called abortion pill, were pursued by people who regarded the use of this drug to be immoral. Efforts to prevent and constrain research involving fetal stem cells also seems to have been pursued by people who regarded this research to be immoral and believed that the method used to obtain these cells was similarly immoral.

The belief that stem cell research is immoral is controversial and debatable. Polls consistently showed that a strong majority of Americans did not agree with this judgment, and regarded this research as important and promising. Nonetheless restrictions on this research were imposed by a presidential order restricting the set of cell lines that could be used in publicly funded research and forbidding the use of public funds for research involving any other lines. Because the available lines were few, and because they had been contaminated with mouse DNA during the process used to "immortalize" them, these restrictions did, at least temporarily, tightly constrain research on human stem cells.

Critics of the Bush administration decision to restrict stem cell research argued that Bush was imposing a parochial moral and religious agenda on others whose convictions were different. A *New York Times* editorial made this case:

> Mr. Bush is adamantly opposed to such research, which involves creating microscopic embryos to derive stem cells that genetically match a diseased patient, thus facilitating research on particular diseases and ultimately potential cures. There, too, he seeks to impose his morality on a society with pluralistic views.[14]

The implicit argument of this passage is that public policy had wrongly been motivated by *private moral* reasons that were not widely shared. In this case, the reasons in question were the president's private religious convictions, which included the conviction that life begins at conception and that fetal life should be protected as sacred. The *Times* urged that it is wrong for the president to appeal to his private moral or religious convictions when enacting public regulations.

The foil to this view, not logically implied but indicated by this writer, is that it would be more proper to permit regulation only when it could be supported by *public* reasons.[15] Where public reasons for regulation cannot be found, people should be left free to do as they wish, without any regulatory impediments. But which reasons are public?

A theory of public reasons is a theory of which reasons justify public action or public regulation. A commitment to public reason involves recognition that public actions must be explained and justified to the people to whom they apply. John Rawls expresses this view as a condition for the proper exercise of power of the government over individuals. He writes, "Our exercise of political power is proper and hence justifiable only when it is exercised in accordance with a constitution the essentials of which all citizens may reasonably be expected to endorse in light of principles and ideals acceptable to them as reasonable and rational."[16] Since regulation involves an exercise of power, we should consider the implications of this strong view for the problem under consideration here. Rawls's requirement for public reason is strong: others should be *reasonably expected to endorse* the underlying principles that justify the exercise of power. They need not agree with the specific exercise of state power in question, but the exercise of power will be unjustified, on this view, if they could not be expected to accept the underlying principle that justifies the use of power. Rawls's view implies recognition that the justification of coercive power is justification *to fellow members of our political community* on whom this power may be exercised, and must be based on principles that they would accept.[17]

In a similar and closely related sense, public reasons may be distinguished from "merely" private reasons that apply to us as individuals, or to members of our smaller private communities. Religious reasons, for example, are "private" in the sense that they apply among fellow believers but would not provide justification for broader public policy. Indeed, religious reasons are sometimes taken as a paradigm case of private reasons that may justify individual action and choice but cannot rightly justify restrictions on the liberty or choices of other people who do not share the same convictions. By contrast, the paradigm examples of reasons that are *public* are (1) constitutional reasons, and (2) reasons that justify regulation or law by reference to our right to protect others from harm.

Perhaps we should provisionally accept the view that scientific research may appropriately be regulated only when the reasons for regulation are based on *public,* not merely private reasons. On this view, wherever restriction or regulation of science cannot be justified by reference to *public reasons,* it is illegitimate and unjustified.

(iii) Public Skepticism about Science

Another motivation for the regulation of science is public skepticism. In the United States, many people are skeptical of scientific results and theories. A significant number of Americans, for example, do not believe the theory of evolution. Many deny the view embodied in the broad consensus among climate scientists, showing the connections between global climate change and anthropogenic greenhouse gas emissions. In both cases, skeptics have made a concerted effort to influence the way science is taught, pursued, funded, and reported.[18]

V. THE LIBERTY TO PURSUE SCIENTIFIC RESEARCH

The discussion above has identified three principal considerations that militate strongly in favor of the liberty to pursue scientific research without impediment. It will be useful to state each of these considerations clearly:

1) Presumption in favor of liberty: The liberty to pursue scientific research is simply one aspect of a more general right against interference from others, at least where our behavior does not threaten harm or risk of harm to others.
2) Freedom of conscience and expression: The liberty to pursue scientific research is implicit in broader protections for other intellectual liberties, including freedom of conscience and free expression.
3) Public reasons requirement: Restrictions on research are only appropriate when they can be supported by *public reasons.* If

regulations are based only on private reasons (the religious or moral convictions of the legislator, for example) they constitute an unacceptable limitation of liberty.

If accepted, these three considerations constitute powerful reasons to avoid many restrictions and regulations that impede scientific research.

VI. PUBLIC FUNDING AND REASONS IN *FAVOR* OF REGULATION

The arguments listed above address *direct* regulation of research: regulations that unconditionally restrict research activities, or which (like IP law) create a regulatory regime that provides incentives. Where research is publicly funded, the case for liberty is different, since there are liberty interests on both sides of the case. Public research funding is effected through taxation, and those who are taxed to support research have a legitimate interest in the research they pay for. In effect, where research is publicly funded, the case one needs to make in order to justify regulation is lighter than it is in the case of privately funded research. Consider the arguments that might be made by a taxpayer whose money is used to support research she might find questionable.

First, such taxation involves both coercion and limitation of liberty, and thus requires justification under the *presumption in favor of liberty* discussed earlier. While direct restrictive regulation of scientific research would require similar justification, the argument from liberty would seem, in this case, to favor the rights of taxpayers and not the rights of scientists and researchers. In this case, if regulations are necessary to ensure that research funding can answer the challenge from the *presumption for liberty*, this would constitute a good argument in favor of regulation.

A second argument from the principle of freedom of conscience also provides support for the regulation of funded research: In general, our right of freedom of conscience protects our right to believe and to express whatever we wish, and is considered to be violated where one is forced to express a view one does not accept.[19] But people frequently take this value to be compromised when their tax dollars are used to support endeavors they do not support. Thus during the war in Iraq, many Americans who disapproved of the war regarded it to be a violation of their right of freedom of conscience that their tax dollars were used to fund a war they did not support. In a similar sense, some people regard it as a violation of conscience that their tax money is used to support research that violates their moral principles. Sometimes this was articulated as an *expressive harm:* "By funding the war with my money, the government forces me to *express* support for a war I do not support." While paying one's taxes is not usually considered to be a fundamentally expressive action, it is easy to understand the view of people who object to their tax money being used in this way. We might call this the "Not with my money!" argument.[20]

But a similar argument arises in the case of controversial research: For example, if federal funds are used to support human fetal stem cell research, taxpayers who are opposed to such research might feel that they are being *forced* to express support for activities they regard as deeply immoral. In this sense, the argument from freedom of conscience provides some significant initial support for the view that this research should not receive public funding. Whether this view is convincing *all things considered* will depend on whether an adequate response can be given to this objection.

A third and closely related argument derives more directly from the requirement that coercive public policies require *public reasons* for their justification. The fact that people disagree about controversial research immediately raises the concern that this expense might not be justifiable *to them* in light of "principles and ideals acceptable to them as reasonable and rational."[21]

Notice that these are the same values we cited earlier in defense of scientific liberty. It seems that the same principles and considerations that support the liberty to pursue research may also provide justification for *constraints* on research that is supported by public funds.

VII. PUBLIC FUNDING, PUBLIC REGULATION, PUBLIC RESPONSIBILITIES

While the considerations cited in the previous section might be thought to call into question the entire institution of publicly funded scientific research, it would be inappropriate to conclude that we should eliminate such funding in an effort to protect the interests of the public. There is a strong public interest in pursuing scientific and technological progress, and without public funding this interest would be poorly served. It is crucial, however, to recognize that the arguments in favor of regulation impose a burden of justification on those who allocate funds for research and those who receive and use them.

Without doubt, some public funds will be used to support research that *some* people will regard as immoral or morally questionable. If universal consent and approval were necessary, then it would be difficult or impossible to justify the public support of any science at all. Since we *do* have good reasons to provide public support for science, we must conclude that the requirement of universal consent is simply excessive. But this conclusion comes at a cost that must be counted: in this case, the cost is borne by citizens who are compelled, through taxation, to provide funds for research they do not understand and of which they may not approve. I would argue that this justifies reasonable public regulation of funded research, and that it also imposes an important obligation on scientists whose research receives this funding. The obligation in question is not simply an obligation to do good science—it goes without saying that those who receive public support have an obligation to deliver quality. But in addition, there is an obligation to do what one can to ensure that the projects one pursues really do serve the public interest in the end and to do what one can to explain one's research to the public whose tax dollars make it possible. Scientists need to be able to explain the value of their work and to show that it really does merit the use of public resources.

IIX. CONCLUSION

This paper began with two pointed questions:

> *What business do nonscientists have to regulate science?*
>
> *What legitimate interest (if any) do legislators and the public have to restrict and regulate the professional activities of scientists engaged in research?*

When considering regulation in the abstract, it may seem that there are overwhelming reasons to leave scientists alone, as free as possible from regulations that might impede or constrain the process of inquiry. Beyond the minimal restrictions necessary to ensure the integrity of the research process and to protect the rights and interests of those who might be harmed or put at risk by some research, scientists should be free to do as they please.

But where public funds are provided to support research, I have argued that researchers should hold themselves to a higher standard, and the case against regulation is weaker. In effect, the burden imposed by the presumption in favor of liberty falls on scientists, not on regulators. Scientists who are granted public support must, through their research, be able to show that the value of their work justifies the infringement of liberty involved in gathering public funds. Even so, not just any reasons will justify the regulation of scientific practice. We should still avoid regulation that cannot be supported by good and sufficient public reasons.

I have argued that the burden is on *scientists* to defend the value of their research and their claim to public funding. Since public funding relies on taxation, the presumption in favor of liberty imposes the heaviest burden of proof on scientists, not on regulators. Scientists who hope to avoid inappropriate regulations had better be prepared to explain their work, and its value, to those who are compelled to support it.

6.

EFFICIENCY VS. EQUITY

Economic Considerations in the Science Debates

Richard Barrett

This essay concerns how economics can be, should be, and is brought to bear on debates about the public implications of scientific developments. Since the role of economics in such debates is not much different from its role in debates about issues *without* much scientific content (welfare reform, say, or globalization), the examples presented here, for what they are worth, are fairly generic. Where possible, the argument will be illustrated with issues of scientific interest, but in some instances an example from a different field will be more useful.

We start by considering the rather extraordinary claim, popular in the halls of business schools and economics departments, that "economics explains everything."

Now nobody in his or her right mind believes this. Indeed, physical and biological scientists would be rather more likely to accept the propositions that "string theory explains everything" or "evolutionary theory explains everything" or "DNA explains everything."

But nevertheless it is instructive to examine what it is that people are apparently claiming when they say that "economics explains everything." Understanding the claim may shed some light on the role of economics in public discourse about contested issues.

Even the most convinced proponent of economics will admit that the "everything" that gets explained is everything in the *social* realm. Economics cannot explain the extinction of tyrannosaurus rex, or what dark matter is, or why the snows are disappearing from Kilimanjaro. So practitioners of the *physical* sciences can feel confident that an economist is not going to come poaching on their professional territory any time soon. But to say that economics explains everything *social*—and to brush aside the political scientists, sociologists, philosophers, historians, and so forth in the process—still seems to amount to breathtaking intellectual imperialism. What's going on here?

It's this: economists believe (or at least assume) that all of us, that is, all people, in all aspects of our lives, are rational, maximizing agents; in other words, we are always making decisions that to the greatest extent possible will make us better off. This type of behavior can be modeled with simple mathematics, and much of economic theory consists of such modeling, although to bamboozle the innocent and impress the cognoscenti, it's better to make the mathematics as complicated as possible.

Formally, agents are said to maximize *utility*, which can be understood as *well-being*, or *happiness*, or something of the sort. Originally, utility was thought of as something *cardinally* quantifiable—"This dill pickle gives me 4 units of utility"—but one of the intellectual achievements of twentieth-century economists was to recast utility theory in *ordinal* terms; this means that agents need only to be able to express a preference between things—"I prefer this dill pickle to that empanada"—and that utility numbers are ordinal; that is, they express preference orderings. In this context, a gain in utility means simply moving to a more preferred state.

No matter how it's conceptualized, utility has two important characteristics. One is that utility is inherently *subjective*, so that it is something you *experience* and the experience is unknown and unknowable to all others, who are accordingly unable to second guess your judgments about what you think is good for you. For many or even most observers, this aspect of utility theory is very attractive; indeed, almost none of them seem willing to turn over their money to someone else on the expectation that these second parties could spend it better on their behalf than they could themselves.

A second, and for many people, less plausible, aspect of utility is that it is undifferentiated by source; that is, things that we buy in markets—movies, pork chops, appendectomies, and so forth—and things that we come by in other ways—love, self-respect, experience of the world—all give the same sort of utility and can be compared for the amount of utility they give. This implies, for example, that parents can actually state the number of, say, pork chop dinners that would give them more utility than, and that therefore they would prefer to, the company of their children. This may be more than they can possibly eat or want to eat in their lifetimes (one certainly hopes so), but the point is that they can entertain the question. Or so the economist supposes.

The subjectivity and undifferentiatedness of utility have important implications to be considered below, but first consider what *maximization* of utility involves. Every choice we make involves trade-offs: when we choose something we gain A and give up B. In economic life, it's pretty obvious: if you want to spend a dollar more on food, you must spend a dollar less on rent, or not save a dollar that could have bought food in the future, or earn the additional food dollar by working more and giving up precious leisure time. Maximizing utility involves constantly assessing these trade-offs, searching for and taking advantage of the opportunity to gain more than you lose. Economists believe that people do this regularly and are quite successful at it; ultimately they find and exploit such opportunities until no more are available; at that point utility is maximized. There is no feasible state of affairs that would be preferable to the one attained.

Subjectivity of utility implies that we cannot know how much utility an individual derives from something. But in economic life we regularly see people buying things, an iPad® or a John Grisham potboiler, for example, and giving good money to do so. But if they are maximizers (which of course we assume them to be) they obviously are buying those things because they expect to get at least as much utility from them as they would get from spending the money on other things. So an individual's willingness to pay for something is taken to be an indication of the utility she expects to derive from its enjoyment or, in other words, the strength of her preference for it. This turns out to be very important.

The undifferentiatedness of utility has several implications that are important for this discussion. One is that people deal with choices in political, social, and personal realms in the same way they do with choices that are explicitly economic. For example, in voting people seek to maximize their personal utility. This, of course, is what leads to the claim that economics explains everything; everything, it turns out, is all choice-making behavior.

A second important implication is that noncommercial things have economic value. We may not have to pay anything to fall in love, but if forced, we would be willing to, and that gives falling in love economic value (that is, it confers utility) just like a donut from Krispy Kreme®. This may seem rather strange, but it does suggest that economists think of economic value as a much different and more expansive concept than commercial value, and it places them in a position of finding economic value in the outcomes of public decisions that most people would never think was there. To an economist, climate stability or uncluttered vistas or the presence of wolves in the Northern Rockies are all economically valuable, and society is justified in incurring costs to capture that value.

This account of the rational, maximizing agent is subject to a variety of objections. For one thing, the subjectivity of utility makes the theory sound dangerously tautological. Consider this dialogue:

> Curious Observer: Why did Joe choose that cheeseburger?
> Economist: Because it will make him better off!
> Curious Observer: How do you know it will make him better off?
> Economist: Because he chose it!

By Now Perplexed Observer: Uh, so he chose it because he chose it??

For another thing, maximizing utility probably doesn't sound like anything any real person actually does when he goes to the store or otherwise makes choices. The data and computational resources required are simply too great. Some scholars, known as behavioral economists, have examined actual decision-making behavior empirically, and utility maximization rarely fares well in the outcome. In 2002, Daniel Kahneman, a psychologist, won the Nobel Prize in Economics for a very large body of work that tends to demonstrate that rational maximization is seriously circumscribed.

Yet another problem is figuring out just where preferences come from: why is it that we believe what we do about the ability of things to make us better off. Economists typically assume that the preference function is "just there," without worrying about where it comes from and how it is formed and altered by experience, culture, technology, and so forth. But of course the utility function is not just there: we learn our preferences from social influences and norms, technology and experience; and Kahneman, again, has reported experimental evidence that this learning doesn't go well. Subjects consistently misremember the pleasure of events (actually, the experiments dealt with pain, but Kahneman assumes memories of pleasure and pain behave symmetrically), and therefore can be expected to derive a dysfunctional set of preferences from experience. Subjects also routinely report that the pleasure they derive from goods tends to match their expectations when the goods are first acquired, but then begins to fade (this has been called the "hedonic treadmill").

Albert Borgmann has argued that technology leads us away from the kind of engagement of our minds and energies that makes for the good life, and into sterile consumption: we don't cook dinner, we open a bag of chips; we don't look at the mountains around us or the moon and stars above us to tell us where we are, we look at a GPS; at the bottom of the Grand Canyon, we don't listen to the canyon wrens, we call our broker on the satellite phone. Borgmann's argument implies that utility maximization is an illusion because technology and culture so distort our preferences and desires, which become so destructive and dysfunctional, that they no longer serve as a reliable guide to making decisions that will really benefit us.

Finally, rational maximization requires that people consider trade-offs that they are often flat-out unwilling to contemplate. There may be parents who would be willing to trade their kids for a sufficient number of pork chop dinners, but most, of course, would never consider it.

What then are the implications when it comes to defining the role of economics in public debates about scientific issues?

First of all, if the benefit (or utility) we derive from the goods and services we acquire through markets determines how much we are willing to pay for them, and if our willingness to pay for goods and services constitutes the incentive for firms to acquire the resources to produce them, it follows that in a wide variety of situations, resources will be guided into the production of things that provide us with the greatest possible material benefits. Another way of saying it is that markets are *efficient*: they generate very high outputs (benefit) per unit of input (resources). And economists love efficiency.

What does this have to do with public debates about science? Many economists would argue that there are some issues that are not really worthy of public debates at all. In their view, all that is at stake in these cases is efficiency, and efficiency will be best achieved by individuals acting on their own in markets.

Consider this example: for some time now, health analysts have engaged in a lively debate concerning whether or not older men should submit to an annual prostate-specific antigen (PSA) exam for prostate cancer. The benefit of doing so is that if undetected, prostate cancer will cause some men to die much sooner than they otherwise would. On the other hand, a large share of prostate cancers will not have this result; indeed, a majority of men over 80 who die of something else will have prostate cancer, and many will have had it for a good long while. So the PSA test is going to reveal a lot of cancers that really would not be a problem, and lead to a lot of cancer treatments that really aren't necessary. But so what? Why take a chance? The answer is that the treatment is expensive and disturbing side effects are common. So is PSA testing a good idea or not? Should there be a public debate among physicians and medical scientists about this question? An economist would probably say there should not be. What scientists should do is make sure that men have the best information possible about the risks and outcomes and then let those men make their own subjective evaluations of those risks and outcomes, and decide whether they want the test or not. No one can second guess these decisions and there's no need for a public resolution of the issue. Incidentally, according to the US Preventive Services Task Force, men over 75 should not routinely undergo the PSA.

Now one might object to the example on several grounds, but the point still stands. In public debates about scientific issues, or any issue for that matter, economists will insist that we should avoid substituting our judgments about what is good for people for their own judgments. Economists take seriously the Latin adage *de gustibus non est diputandum*—there's no disputing tastes. Now we can all agree that for many reasons we should *of course* respect the autonomy of individuals and should *of course* not barge into people's lives telling them what is good for them, but it does happen.

Economists regard this substitution of judgment as inappropriate because they want public decisions to be efficient, which can only be the case when they conform to the needs and preferences of the affected parties as they themselves express them, supposing the affected parties to be competent adults. But returning to Albert Borgmann's argument, if competent adults' preferences are demonstrably dysfunctional, then it may well be that someone knows what's good for some people better than they do themselves, and that knowledge should be brought to bear in public debate, at least on efficiency grounds. On other grounds, such as respect for the autonomy and privacy of individuals, it might still be excluded.

Consider a concrete example: regulating snowmobile use in Yellowstone National Park. This is clearly a controversial issue, and economic, social, and ecologic arguments for both allowing and banning snowmobile use abound. Proponents of a ban occasionally offer as an argument that snowmobiling is not good for the people who do it; that they would engage their minds and bodies in the experience of Yellowstone in the winter if they got off their high-speed, noisy, distracting machines and stepped onto a pair of skis or snowshoes instead. Does this argument belong in public debate? A convinced economist would reject it out of hand, but Borgmann might not.

Leaving aside the economist's touching faith in the infinite wisdom of the rational, maximizing agent, a reason not to make arguments of this judgment-substituting type is that they are bound to be perceived as patronizing and insulting by people who ride snowmobiles, and patronization and insult are corrosive in the process of deliberating about difficult and controversial public issues. But then again, Borgmann's argument suggests that there may be a great deal at stake here, and we omit its consideration at our peril.

This is not a trivial matter. Arguments that involve the substitution of judgments are not just an occasional curiosity in public debates, and if we want to be skilled and effective deliberators, we need to know how we should regard them.

So markets are great, aren't they? Well, for some things they are pretty good, but economists have also described a variety of common circumstances in which they fail, that is, in which they do not provide for the capture of net gains. There's something (cleaner air, say) that's worth a great deal to us and which we could attain at minimal cost (by installing pollution control equipment). Cleaning up the air would therefore result in significant net gains, but there is no market that will make it happen. Collective action is required. So here is an important point: *from the point of view of economics, the purpose of, and justification for, collective action is to capture net economic gains; that is, to enhance the efficiency with which we use and manage our resources.* And when it comes to debating alternative courses of action with respect to climate change, or biotechnology or nanotechnology policies, or whatever, the preferred alternative is the one that enhances efficiency the most. Indeed, left to their own devices, economists would resolve critical public debates by appealing to the efficiency characteristics of the alternatives.

There is an interesting twist on this argument. Some time ago Ronald Coase, a Nobel laureate in economics, wrote a very influential paper titled "The Problem of Social Cost," in which he argued that market failure is the result of high transactions costs. Consider the air pollution example: the value to the public of cleaning up the air is greater than the cost to the polluter of installing pollution control equipment. Why doesn't the public pay the polluter to clean up? Because the thousands and thousands of individual transactions that would have to occur would each be costly but individually would capture a very small share of the total benefit to be realized. So it doesn't happen. The philosopher Carl Sagoff has pointed out that what Coase's argument really implies is that there is no such thing as market failure at all. If the benefits of cleaning up the air don't outweigh the combined cost of the pollution control equipment and the transactions required to get that equipment installed, efficiency requires that the air not be cleaned up. The market has done the efficient thing and has therefore not failed. So, Sagoff says, if markets can't fail, there is no economic justification for collective action, and considerations of efficiency can have no bearing on public debates, which must therefore be resolved on other terms.

What can we say about efficiency as a value in debating scientific issues? Well, for one thing, it should not be dismissed as trivial or irrelevant in the light of our larger purposes. In a society as affluent as ours, it appears unseemly, even greedy, to worry about squeezing ever more out of our resources. But there are plenty of places around the world where becoming more efficient by improving the productivity of land, of resources, and of people may mean the difference between surviving and not. That surely is not something we should lose sight of when we debate, for example, the international distribution of genetically modified crops.

We can come at this point from another direction. Consider policies intended to reduce carbon emissions in compliance with international agreements. Broadly speaking, there are two major approaches to the design of such policies. One involves the mandating and/or encouraging of specific carbon-lowering strategies such as the production of biofuels and wind energy, higher mandatory fuel efficiency standards for vehicles, mandatory sequestration of carbon in conventionally fueled power plants, and so forth. The other involves the capping of carbon emissions and permitting the trading of emissions allowances among capped sources of emissions, so that a source that emits less than its cap can sell the difference to another that will emit more than its cap. There are all kinds of reasons to like or dislike these alternative approaches, and surely we have an occasion here to debate. An economist entering the debate would say that the second policy—cap and trade—is highly preferable on efficiency grounds. The reason is that under cap and trade, emissions reductions will be undertaken by those who can do so most efficiently and at least cost (because then they can make money selling their excess allowances), while there is very little prospect that the mix of strategies called for under the first, more directive, policy would be similarly optimal.

However we do it, significantly reducing carbon emissions is going to be costly, indeed so costly that so far the United States Senate has been unable pass a bill to do so. Of course there are issues other than cost that create the Senate logjam, but if cost is going to make or break our response to climate change, then we should listen to the economists carefully. Unless we have efficient policies, we may have no policies at all. Cap and trade, of course, is no substitute for political will. The European Union installed a carbon cap and trade system, but then was unable to bring itself to set the caps low enough to make any difference. So in that case despite having an efficient policy, in effect they still had no policy at all.

The application of efficiency analysis to the design of emissions reduction policies is fairly straight forward. It assumes that we have already decided that *whatever it costs*, reducing emissions is worth it. Being efficient in this context simply means getting the costs as low as possible. But if they had their way, economists would probably prefer to formulate the problem differently: They would ask whether the benefits of reducing emissions (which are the avoided costs of uncontrolled climate change) outweigh the control costs, or perhaps whether it would be more efficient to let emissions continue as they otherwise would, allow climate change to occur as it otherwise would, and adapt to it. What economists are asking us to do here is strike a "reasonable" (which to economists means "efficient") balance between adapting to and arresting climate change.

Now every instinct you have may tell you to reject this balancing act: surely this madness must stop. But it's too late. Adaptation is far from hypothetical. Climate change is occurring and we are adapting; climate change is going to continue to occur and we are going to have to continue to adapt. Can cost-benefit analysis tell us how far down this road we should go? Well maybe. But there are some serious problems.

One is that while the idea of comparing the economic magnitudes of costs and benefits is sensible enough, when it actually comes to establishing those magnitudes, all bets are off. What dollar sign can we put on the disappearance of the Artic sea ice? What about the loss of a way of life for the indigenous peoples who hunt on that ice or the polar bears that live on it? What is the value of what will be lost as Miami slowly goes under water? Rising water temperatures and diminished in-stream flow threaten trout in Montana. What is the value of a healthy trout population? And so forth. It turns out that fools rush in, and economists have indeed developed many techniques intended to establish these kinds of values (remember: willingness to pay is key here), but the results are discouraging in their implausibility, variability, and inconsistency across measuring techniques. More to the point: relatively few people involved in the debate about how we should respond to climate change regard these measures of value as valid or legitimate.

It is also difficult to conduct a cost-benefit analysis of alternative strategies for responding to climate change when there is so much uncertainty about what is going to happen, when decisions can have irreversible consequences, and when many of the parties affected by the decisions we make today are not around to express their preferences (or willingness to pay). It's true that economists have devised a variety of ways to deal with all these flies in the ointment, but their applicability to the climate change problem at hand is not very robust.

A final, and perhaps the most telling, objection to applying cost-benefit analysis in public debates is that in some circumstances it is inappropriate or irrelevant. This is because there are other values, principles, or convictions involved in deciding the issue that trump efficiency. Carl Sagoff makes this point forcefully with respect to the Endangered Species Act (ESA). Why, Sagoff asks, do we require government to do whatever it must to protect endangered species, almost without regard to the cost of doing so? Is it because we know that the economic value of the continued existence of a species is necessarily greater that the cost of protecting it? To Sagoff, the answer is pretty clearly no; we protect species because we have an obligation to do so (it is the ethicists who must explain where that obligation comes from), and the obligation doesn't cease to exist just because it's costly to live with.

When it is appropriate to consider efficiency and when it is not is an interesting problem in debating public issues. There are clear-cut instances in which decisions should be made mostly or entirely on efficiency grounds. Should we produce lettuce or spinach? Well for Pete's sake, produce whatever it is that customers want and are willing to pay for. At the other end of the scale there are cases, like Sagoff's example of endangered species, where efficiency shouldn't count at all. It doesn't make any difference how much people are willing to pay for a round of golf, we are just not going to destroy grizzly bear habitat to build a golf course on it. Notice, by the way, that there is no unanimity on this last point. There are plenty of golf course developers who would like to see the ESA itself become extinct.

It's in cases between these two extremes that we need to figure out whether economics is relevant to the debates before us. Consider this one: about the time that *An Inconvenient Truth* was winning its Oscar, it was revealed that Al Gore was using an awful lot of energy to heat and light his great big house in Tennessee. Gore defended himself by announcing that he was purchasing carbon offsets, meaning that for every pound of carbon that he was responsible for putting in the atmosphere, he paid somebody, somewhere, to take a pound of carbon out. To an economist, this was the sensible, efficient thing to do. Strictly voluntarily, Gore wanted to reduce his carbon footprint to zero. He had two ways to do it: turn off all his lights or pay somebody else to plant trees or build a wind farm or what have you. He chose the second way because he regarded it as less costly. What can possibly be wrong with that? Well, according to Charles Krauthammer, writing in *Time* magazine, Gore was simply buying an indulgence, forgiveness for the sin of excess carbon emissions. Even Michael Kingsley, again in *Time*, posed the question that if it's okay to buy an offset for carbon emissions, what about a child abuse offset? Obviously if you think that emitting carbon is the moral equivalent of beating up a child, you aren't going to find cost-benefit analysis of much use in dealing with the problem.

There are all sorts of values, imperatives, convictions, and obligations that might supersede efficiency as a consideration in public debates, but among these the one about which economics might be expected to have the most to say is *equity*, that is, the question of who shall bear the costs and who shall enjoy the benefits of the decisions we make. What can economists say about equity?

Almost every collective choice we make that alters the economy in some significant way creates winners and losers. Economists specifically disavow any special knowledge about whether the pattern of gains and losses is appropriate, fair or just; take that question down the hall to the philosophy department. More importantly, given that utility is entirely subjective, there's no way of comparing it from one person to the next, which means that it is impossible to say if a particular redistribution of income is efficient or not. Even if we take $100 from Bill Gates (who wouldn't seem to need it) and give it to an impoverished graduate student (who's subsisting on peanut butter and jelly sandwiches), we don't know whether the total utility experienced by the two of them will rise or fall.

That may seem like a rather serious limitation, but consider this corollary: to know that a collective decision we have made is going to raise total utility, that is, is going to enhance efficiency, there either have to be no losers or the winners have to be able to fully compensate the losers and have something left over for themselves. In other words, any efficient decision is potentially equitable, provided that the appropriate mechanisms for paying compensation are in place.

In practice, of course, compensation is rarely forthcoming. We see it occasionally, as when, in conjunction with entering into free trade agreements, we establish assistance programs for workers displaced by cheaper imported goods. But usually compensation remains little more than a twinkle in the eye of the winners; when, under the North American Free Trade Agreement (NAFTA), imports of US corn wiped out impoverished corn producers in central and southern Mexico, farmers in Iowa and Mexican urban consumers weren't sending checks to their counterparts in Chiapas. Efficient choices can then lead to unjust outcomes, and when they do, we are often inclined to make a less efficient choice with a fairer distribution of consequences. Indeed, in the debates about the economics of global warming, or globalization, we hear a great deal about who is going to benefit and who is going to suffer, but very little about efficiency. It's interesting that implicit in these arguments is the assumption that "more is better," but little attention is paid to efficiency, which is the most important way of getting more.

One important way in which economists can participate in discussions about equity is in clarifying who the winners and losers really are. In this regard, things are seldom what they seem. We have a tendency to assess the results of our decisions by focusing on their proximal impacts and ignore how those impacts are suffused throughout the economy in a network of markets and transactions. For example, when we require coal-fired power plants to sequester carbon dioxide, we are inclined to believe that it is the power company that will bear the costs, which seems fair. We tend to ignore the fact that the company will succeed in shifting a good deal of the burden to its customers, or to the ranchers from whom they lease the coal, or the railroad that transports the coal, or to the workers in the power plant. Figuring out all of these effects is not easy, but we want to bring them into the debate (and we should); we should bring an economist into the debate as well.

Another aspect of the discussion of equity to be borne in mind is what economists call *rent seeking*. This is when private interests go to government and the community at large seeking arrangements that will raise their incomes. For example, a pharmaceutical company seeks to extend its patent over a drug or secure the cooperation of the government in enforcing its patent rights in other countries—AIDs drugs in Africa, perhaps. Now nobody ever admits to rent seeking: "Please do this so I can get richer." Rather, they represent the arrangement they seek as serving some public interest, including equity. The rent they seek is represented as something that by community standards they deserve as a matter of fairness. Sometimes, of course, they do, but separating the wheat from the chaff in these claims is tough going. In public discourse on these kinds of issues, the lines between patently special interests, genuinely public interests, and the attainment of equity are easily blurred by those who benefit from blurring them.

This raises a final point: quite often—way too often really—the economics that's brought to bear on public debates consists of economic impact analysis rather than efficiency or equity analysis. In impact analysis, some variable is chosen (usually it's jobs or wages), the impacts of different alternatives on that variable are calculated, and the alternative with the biggest impact is preferred. Should a roadless area be designated wilderness or logged? Well, let's see: are there more jobs in logging or in outfitting? Ah, logging. Well there's the answer. Crude but amazingly common.

The first thing to be said about impact analysis is that it is almost always wrong, with impacts that are grotesquely exaggerated. Grotesque in this context means roughly one order of magnitude.

The second thing to notice about impact analysis is that it dresses itself up as cost-benefit analysis because impacts (more jobs, for example) are assumed to be good things (benefits). The lunacy of this association with cost-benefit analysis is not hard to discern. Across the West every summer, and particularly in California, hundreds of workers have risked their health and lives to fight wildfires, which are arguably more severe due to climate change. So climate change creates jobs, and that's good for us. In cost-benefit analysis jobs are costs, not benefits. If a farmer asked us whether planting wheat or sunflower seeds represented the highest and best use of his land, would we tell him to plant the one that takes the most work?

Is "growing the economy" really a benefit in a community whose principal preoccupation is the costs of growth?

Why are impacts treated as though they are synonymous with benefits? More often than not, it's because the impact estimates are being produced by rent seekers who have a vested interest in blurring the line between what is good for them and what is good for the larger community. That also explains why the impacts are grossly exaggerated.

Impact analysis is probably the most common and well received way that economics is brought to bear on public debates and decision making, which is odd because it is also the line of analysis that is least grounded in the fundamental structure of economic theory described at the beginning of this essay.

In the end, then, economics doesn't explain everything, and some of what it does explain is beside the point. The challenge for those engaged in the discussion of public issues is to find what is useful and legitimate in economic analysis and to recognize and reject its misuse.

7.

COMMUNICATING SCIENCE

Moral Responsibility in Theory and Practice

Wendy S. Parker

1. INTRODUCTION

A climate scientist examines new results from a computer simulation study. Let us suppose that, according to those results, if greenhouse gas concentrations continue to rise at current rates, the next century will bring substantially drier conditions in several parts of the world, including regions where serious human suffering already occurs in times of drought. Suppose further that these new results are in agreement with simulation results obtained by other research groups and are also supported by physical understanding of the processes that influence how much rain occurs in these regions. Considering all of this, the climate scientist deems it likely that, if greenhouse gas emissions continue rising at current rates, the next century really will bring drier conditions and more frequent droughts in these regions.

What should this climate scientist do? Is it enough for her to report her simulation results in a scientific journal? Or should she contact newspapers and other media outlets to try to convey to the public her expectations about the future? Should she go even further and become an advocate for policy action to reduce greenhouse gas emissions or find alternative water sources for the drought-prone regions? This paper begins to explore questions like these about the moral responsibilities of scientists.

In section 2, I offer a simple moral argument for the conclusion that, when their research findings indicate a serious threat to humans, scientists ought to communicate those findings (and the associated threat) to the public. This argument—only one of several that might be given for the same conclusion—relies on the idea that when people are made aware of potential harms to themselves or others they are often in a better position to avoid or otherwise mitigate those harms. But the simple argument says only that communication should occur, not what it will amount to in practice. And in practice, communication can be complicated by a number of factors. In section 3, I discuss four of these factors—*uncertainty, multidisciplinarity, complexity,* and *politicization,* illustrating the challenges they bring when it comes to communicating climate change findings like the hypothetical ones mentioned above. Finally, section 4 offers some concluding remarks.

2. A SIMPLE MORAL ARGUMENT

Let us construct a simple argument regarding scientists' obligations to communicate what they learn. The first premise of the argument will be a general principle: (P1) *If an action by a moral agent will help prevent serious harms to humans, and if performing the action would not involve sacrificing anything of moral significance, then the moral agent ought to perform the action.* P1 is inspired by the "moderate principle" offered by utilitarian philosopher Peter Singer, for whom sacrificing something of moral significance means (roughly) either doing something that is morally wrong in itself, causing some significant harm, or failing to promote some important good that one is otherwise expected to promote.[1] P1 suggests that if we see an injured pedestrian in urgent need of medical care, and we have a cell phone that we could easily use to call an ambulance, then we ought to do so, rather than just pass by. It does not require, however, that we starve our own children in order to feed the children of others. P1 is offered here as an intuitively acceptable moral principle that might be defended within various frameworks (not just a utilitarian framework).

The second premise of the argument will be: (P2) *Scientists are moral agents.* This premise is difficult to dispute, since scientists are intellectually competent individuals, able to plan and choose courses of action. It is sometimes claimed that scientists fall into a special category of moral agent, excused from ordinary moral responsibilities for the sake of scientific progress. However, this has been forcefully challenged by philosophers.[2] Moreover, in practice there is evidence that moral considerations *are* believed to trump enhanced opportunities for discovery. For instance, when scientists experiment on human subjects they are required to obtain informed consent and to follow other rules that help protect the subjects, and some experiments are deemed simply off limits. The burden of proof seems to be with those who would deny P2.

The third and final premise of the argument will be: (P3) *In many cases, when scientific research findings indicate a serious threat to humans, (a) communicating those findings (and the associated threat) to the public can help to prevent some of the threatened harms, and (b) such communication would not involve sacrificing anything of moral significance.* The justification for (a) is the commonsense idea that, in many cases, when people are made aware of potential dangers, they are in a better position to avoid or mitigate them. For instance, understanding that smoking causes cancer deters some people from starting to smoke and leads others to convince loved ones to stop, avoiding some cancers that would otherwise occur. Likewise, when warning sirens communicate that there is a tornado in the vicinity, some people move to safer locations; this too can help to prevent some injuries that would otherwise occur. The justification of (b) comes via simple reflection on scientific practice: typically, communicating research findings (and associated threats) to the public is neither morally wrong in itself, likely to cause significant harm, nor done at the expense of some important good that the scientist is otherwise expected to promote. Exceptions can arise, of course. For instance, if communicating one's research findings can be expected to instigate mass panic that in turn can be expected to result in numerous accidental deaths, then (b) might fail to hold.[3]

Together, P1–P3 warrant the following conclusion: *In many cases, when their research findings indicate a serious threat to humans, scientists ought to communicate those findings (and the associated threat) to the public.* So here we see a simple moral argument—ultimately underwritten by the idea that harm to humans is a bad thing—that calls for scientists to communicate at least some research findings to the public. Other arguments to a similar conclusion are clearly possible. For instance, we might focus on the fact that a great deal of scientific research is publicly funded, arguing that this creates an obligation on the part of scientists who receive such funding to inform their fellow citizens of relevant findings. Moreover, since scientists may be the only people with the knowledge and expertise needed to foresee some threats, it might be argued that it would be particularly bad for them to fail to communicate those threats. But setting aside such additional arguments for now, let us suppose that the simple moral argument given above is persuasive on its own.

3. BEYOND THE SIMPLE ARGUMENT

The simple argument says only that communication should occur, not what it will amount to in practice. And in practice, a number of complications can arise: *uncertainty* can make it difficult to determine when claims should be made; *multidisciplinarity* can render it less clear what should be communicated by whom; the *complexity* of scientific issues can make it hard to see how findings should be communicated (i.e., using which terminology, including which caveats, etc.); and the *politicization* of scientific issues can complicate all of these dimensions of communication. In what follows, each of these is briefly discussed and then illustrated in the context of communicating about climate change.

3.1 Uncertainty

Science is always open to revision; even today's best scientific theories will probably turn out to be mistaken in various ways. But at any given time, there are some questions that science seems to have answered rather definitively, while for other questions the answers are still emerging or even missing completely. For example, there is little uncertainty about whether the Earth is flat, but there is significant uncertainty about the number of other planets in the universe that are home to intelligent life. Indeed, we can imagine a spectrum of scientific questions, ranging from those whose answers remain highly uncertain to those for which answers can be confidently given because the available evidence is overwhelming, with many questions of interest falling somewhere in between.

In the case of climate change, there is now substantial evidence that Earth's climate is warming on a global scale and that greenhouse gas emissions related to human activities are the main cause of this warming.[4] By contrast, considerable uncertainty remains about how climate will change over the coming century under different greenhouse gas emission scenarios, especially when it comes to regional and local changes.[5]

That is, for many regions of the world, the scientific situation is not as clear as in the hypothetical example given at the start of this paper—simulation studies may not all agree in their results, and physical understanding on its own may be insufficient to determine even the expected direction of change for some conditions (e.g., wetter or drier).

So if a few simulation studies indicate that the next century will bring substantially drier conditions in impoverished (and already dry) regions, while other simulation studies show little change in future rainfall for these regions, what should scientists communicate to the public, if anything? Should they warn that increased drought is quite plausible, even if they cannot say how likely it is, or should they wait until there is less uncertainty about what will happen? More generally, should every scientific result that portends serious harm to humans be communicated to the public? How strong does the evidence have to be that significant harms will occur (or are occurring) before scientists should communicate the threat? This is the challenge of uncertainty.

Philosopher Heather Douglas has argued that, when deciding whether to make a claim, scientists ought to consider the consequences of being mistaken. That is, scientists ought to consider what will happen if they assert that X is the case, but it turns out that they are wrong; to the extent that such a mistake would have very negative consequences, scientists should require a high standard of evidence (relatively low uncertainty) before asserting X.[6] For example, before a scientist claims that common antibiotics can occasionally cause cancer in humans, he should consider the consequences of being wrong about this. Perhaps he can foresee that, in light of his announcement, many people would choose not to take antibiotics when they are ill, so that if he is wrong, there will be unnecessary suffering and deaths from untreated illness—perhaps even more suffering and death than would be expected to occur, if he is right, via the cancer caused by antibiotics.

As the antibiotics example illustrates, false alarms can lead to harms that otherwise would not have occurred. False alarms can also damage scientific credibility, reducing the effectiveness of later warnings about threats that are real. In fact, even if it turns out that a threat is real, credibility can be damaged if it is revealed that the alarm was sounded when the evidence clearly was not sufficient.

When it comes to our climate change example, Douglas's view would counsel that, before claiming that various drought-sensitive regions will be even drier in the future, scientists ought to consider what consequences can be expected if their claim turns out to be a false alarm. But while the consequences of being wrong about the antibiotics were perhaps not so difficult to foresee, they seem less clear in the climate case. So far, when threats related to climate change have been announced, little action has been taken to mitigate or prepare for those threats. Will the same be true in the case of the drought findings, or will this announcement finally catalyze action? If actions are taken, what will they be, and will they have harmful side effects that otherwise would not occur? What will be the cost to scientific credibility if it is revealed that a false alarm was sounded in this case? The answers to most of these questions seem far from obvious.[7]

3.2 Multidisciplinarity

A second source of complication is the fact that many issues of interest to the public (and to decision makers in government and industry) span multiple scientific disciplines. A conclusion about future climate change is typically the endpoint of a long and complex chain of inferences that draw upon information from a wide variety of scientific disciplines, including atmospheric physics and chemistry, hydrology, oceanography, and sometimes ecology and plant biology. Detailed conclusions about the harms that climatic changes can be expected to bring to humans may rest, in addition, on information from the fields of medicine, agriculture, and economics.

Can any individual have expertise broad and deep enough to competently assemble and/or assess these long and complex chains of inference? Should an expert in a given subfield (e.g., tropical oceanography) communicate to the public the conclusions that he as an individual has reached about future climate change, or will he be stepping too far beyond his own expertise? Should only climate change conclusions reached by multidisciplinary expert groups be communicated to the public? Multidisciplinarity can make it less clear whose conclusions about expected harms (or lack thereof) should be communicated, and by whom.

3.3 Complexity

The complexity of many scientific issues is a third problematic factor. Rarely will it be enough for scientists to simply state (e.g., in a newspaper article or an interview for television) that there is now convincing evidence that X is the case, where X is some alarming hypothesis or conclusion. There will be questions about what the evidence is for X and perhaps also why that evidence is deemed sufficient for believing or accepting that X is the case. For instance, climate scientists cannot simply report that humans are causing Earth's climate to warm and that even more global warming is expected if greenhouse gas emissions continue to rise; they are called upon to further explain what the evidence is for these claims, why this evidence is sufficient, etc. And so some account of the greenhouse effect is required along with—it would seem—a lot more information: about where greenhouse gases come from and how they are measured, about observations of changing climatic conditions at various locations around the globe, about how heat is moved within and beyond the climate system, about computer models of the climate system and their results, and about the potential errors and uncertainties associated with all of these.

Yet in mass-media contexts, scientists often have only a few minutes, or a few paragraphs, to try to communicate key conclusions and the evidence for them. Even when more time or more space is available, simplifications and omissions will be required, in part because the "full story" would assume specialized knowledge that most of the audience does not have. Interesting questions arise, however, when we consider what sorts of simplification and omission should be made: is it more important that what is said both covers key conclusions and calls attention to caveats/uncertainties, or that most audience members come away with an accurate picture of the perceived threat and the scientist's (or scientists') judgment of the strength of evidence for its reality?

The first option fits better with the usual norms of science—conclusions are to be reported along with caveats/uncertainties—while the second option makes successful communication of potential harms the top priority. These goals can be in harmony, but they need not be.[8]

For some threats, for instance, typical cognitive biases of human beings might be such that a more accurate sense of those threats will be conveyed if scientists leave out caveats and downplay uncertainties. Most of the very negative impacts of global climate change, for example, will not occur for several decades. Yet people have a tendency to underweight the seriousness of threats whose potential harms are so distant in time.[9] This tendency might be exacerbated if scientists emphasize the uncertainties associated with future climate change but might be compensated for (to some extent) if scientists downplay those uncertainties and draw attention to the scariest of the plausible impacts.[10]

If so, should scientists downplay uncertainties and emphasize scary impacts?

The trouble is that such scientists, even if they were motivated purely by the goal of having members of the public come away with an accurate picture of the perceived threat, would begin to look like policy advocates. And here we arrive at the fourth complication for communication: politicization.

3.4 Politicization

In debates surrounding many social and environmental issues, there is a hair-trigger threshold for accusations of misrepresentation and bias, targeting not just politicians and self-acknowledged advocacy groups but also scientists. Especially if science becomes a focal point in these debates, scientists cannot assume that they will be seen (or portrayed) as well-intentioned and honest. Indeed, when deciding how, when, and what to communicate, it may be wiser to assume that one's words and actions will be characterized in the most unflattering and self-serving light possible. When discourse surrounding scientific research becomes highly politicized, as in the case of climate change science, communicating research findings and associated threats becomes extra challenging, susceptible to being framed as advocacy, even when advocacy is not the goal.

But what's wrong with scientists' advocating for policy action anyway? If a climate scientist, on the basis of her scientific understanding and her personal values, believes that greenhouse gas emissions ought to be drastically reduced and that industrial nations should bear most of the cost, why shouldn't she express this view publicly and encourage the sort of policy action that she deems best? After all, scientists would seem to have as much right as other citizens to express preferences regarding policy action. Yet if science has become a focus of the debate surrounding policy options, as in the case of climate change, then scientists who express such preferences and judgments—even when making clear that they are speaking as concerned citizens and in light of personal values—may be accused of lacking objectivity in their scientific work. This can put scientists in a difficult position, facing a choice between maintaining scientific credibility and pushing for policy actions that they consider urgent and/or morally required.[11]

As a matter of logic, of course, from the mere fact that a scientist supports some policy action on an issue (e.g., water pollution), it does not follow that the scientist's research related to that issue cannot be as objective as that performed by scientists who are politically disengaged. Indeed, it seems unfair to make such an accusation without real evidence that the scientist is no longer acting in accordance with the usual standards and norms of scientific research. At the same time, it is an interesting psychological question whether becoming a policy advocate increases the risk that a scientist will unknowingly discount scientific evidence that fails to support her preferred policy option. Investigating this psychological question is beyond the scope of this discussion, but it is worth noting that there are ways to mitigate the risk of such unintentional bias.

One way to mitigate the risk is for scientists with strong policy preferences to acknowledge to themselves the possibility of unintentional bias and to explicitly remind themselves that one or more of their conclusions might be wrong. A second way, and one which has more hope of reassuring people who are suspicious, is to seek opportunities for open engagement with scientists who interpret the available evidence differently. Ideally, this engagement would not come in the form of a debate where rhetoric and strategic maneuvers are employed to make the "other side" look less credible. In fact, that sort of exchange may only reinforce the public's suspicions that the politically active scientist will readily bend science to support her preferred policy action.

A much better option would be a cooperative dialogue in which all participants not only demonstrate a willingness to understand and take seriously the points of view expressed by others, but also engage with one another in an honest and respectful fashion and remain genuinely open to revising their views (e.g., about the strength of available evidence for a scientific claim) in light of the considerations and arguments presented in the course of the dialogue. Here, scientists would be engaged in a kind of deliberative process, working toward the mutually agreed-upon goal of determining what are the best reasons to believe, rather than in an activity whose ultimate goal is persuasion. The results of such deliberation regarding scientific matters might then inform further deliberation undertaken in the same cooperative spirit, but this time involving members of the public with diverse perspectives and values (without necessarily excluding scientists) and having as its goal the identification of more and less attractive policy options.

4. CONCLUDING REMARKS

The simple moral argument given in section 2 is only a starting point; while calling for the communication of scientific results that indicate a serious threat to humans, it does not specify what that communication should amount to in practice. Using the case of climate change, we have seen that uncertainty, multidisciplinarity, complexity, and politicization are factors that can complicate the communication of scientific results (and associated threats) in practice. Key questions that emerged from a consideration of these factors included:

Uncertainty: How strong should the evidence be that significant harms will occur (or are occurring) before scientists alert the public?

Multidisciplinarity: When evaluating the evidence for a threat requires expertise from several disciplines, should only conclusions reached by multidisciplinary expert groups be communicated to the public?

Complexity: How much scientific accuracy should be sacrificed for the sake of successfully communicating to non-experts the nature of a threat revealed by complex scientific research?

Politicization: As science becomes a focal point in policy debates, how can scientists exercise their rights as citizens—openly advocating or criticizing policy proposals—without losing their scientific credibility in the public eye?

These questions stand in need of further attention. Scientists, philosophers, experts in communication, and the general public would all seem to have something to contribute, whether the goal is to uncover insights with broad applicability or to address just one of the questions in connection with an especially controversial or important issue. Indeed, it may be that the best way to begin to address these questions is via the same sort of cooperative dialogue described at the end of the last section, bringing together different stakeholders—experts and members of the general public alike—with the goal of formulating some preliminary guidelines that can serve as a launching point for further discussion.

Part Two

SCIENCE AND TECHNOLOGY DEBATES

Section 1.
BIOTECHNOLOGY

8.

IS SUSTAINABILITY WORTH DEBATING?

Paul B. Thompson

WHY DEBATE SUSTAINABILITY?

Sustainability is a good thing. Why would we want to debate it? The answer to this question lies both in the story of how sustainability came to be so important, as well as in the need to ensure that science to promote sustainability maintains a critical edge in the future. In fact, there could be several books written under the title "Debating Sustainability." The treatment that is given here is both succinct and selective. I begin with a brief historical orientation to the very idea of sustainability, then sketch the compromise approach that seems to be dominant in discussions on sustainability today. I argue that we can do better than this compromise approach if we recognize that most technical models or specifications of sustainability rely on one of two key concepts. One is resource sufficiency, the idea that resources are foreseeably available. The other is functional integrity, the idea that key systems are resilient in response to external threats. A third notion, which I call critical engagement, emphasizes the need for social processes that surface and ameliorate conflicts arising from repression or domination. Getting at least these three notions into one's head is critical to seeing how others may be approaching sustainability in their science and their action.

HOW WE GOT TO SUSTAINABILITY

When I installed the Microsoft® Word 2003 word processing program on my computer about six or eight years ago, the word "sustainability" was not included in the default spelling dictionary that came with it. When I installed the latest version about two years ago, it was. This obscure difference testifies to the distance that our concept of sustainability has come over a very short time. The adjective "sustainable" has been with us for a long time, as has the verb "to sustain." But as something to be named with a noun, the characteristic of being sustainable is of relatively recent origin. These origins themselves might be debated. At least by the mid-1970s there were starting to be discussions of sustainable agriculture among those who were already becoming convinced that the industrial food system was reliant on too much fossil fuel, was consuming too much water, and was not replenishing soil tilth or fertility. Agricultural economist Lester Brown was a participant in these debates, and some credit him with coining the expression "sustainable development" at about the same time.[1]

Carl Mitcham has traced the recent origins of sustainable development to the book *Limits to Growth* published in 1974, though the book does not use the word "sustainable" in any theoretically precise or important way. He writes that this concept becomes important in two key texts, the *World Conservation Strategy*[2] and *Our Common Future*.[3] The latter was the report of the World Commission on Environment and Development (WCED) chaired by Gro Brundtland and often referred to as "the Brundtland Report." All of these documents were spurred by the growing recognition that depletion of fossil fuel reserves would increasingly constrain the economic growth processes that were characteristic of the twentieth century. In addition, analysts such as Brown recognized that potentially renewable resources such as soil and water were being exploited in ways that undercut the ecological systems on which their replenishment depends. The succinct definition of sustainable development as "development that meets the needs of the present without compromising the ability of future generations to meet their own needs," first gained widespread circulation through the Brundtland report.[4]

The decade following the Brundtland report saw many scientists and analysts from different disciplines scrambling to fill in the broad language of the WECD with more precise economic and biological concepts. Much of this debate revolved around an economic approach that defined sustainability in terms of perpetual economic growth. Advocates of this approach presumed that as scarcity of fossil fuels increased, so would the price of these resources. This would have the effect of incentivizing both conservation and the search for technological alternatives. The risk to sustainability that theorists in this group saw resided primarily in the potential for wealthy people and nations to constrain economic growth among the poor. By monopolizing both the control of resources and the wealth that they generated, powerful nations and peoples would prevent poorer peoples from achieving the economic development that they needed to escape extreme poverty. These economically oriented theorists saw poverty as a greater threat to sustainability than the kind of absolute scarcity of resources that had been noted in *Limits to Growth.* Opponents of this school appeared not to notice its fundamentally liberal and egalitarian orientations, focusing instead on a defense of the *Limits to Growth* view that a decline in the planetary capacity to produce resources represented the greatest challenges to sustainability. They tended to see the economists' focus on growth as a retrograde defense of capitalism, and were intent on promoting the idea that growth has had its day.[5]

There were also a number of attempts to establish consistent approaches within specific disciplines. The American Society of Agronomy, for example, held a symposium on sustainability that became notorious for the acrimony among participants and for the utter lack of any agreed upon framework for discussion.[6] The upshot of this intense debate between 1987 and 1997 was that many people became convinced that sustainability *was not* a useful idea for guiding research or policy in the face of increasing resource scarcity or failing ecosystems.[7] As more and more were disenchanted with the idea, the enthusiasm for linking sustainability to environmental science declined in the final years of the twentieth century. Yet sustainability was not dead. A spate of publications at the turn of the millennium offered a model in which sustainability was described as having three overlapping dimensions.[8] Because these publications often present the concept through three overlapping circles, I refer to this idea as "three-circle sustainability."

The three dimensions are generally referred to as economic, environmental, and social. Sometimes the concept is expressed as "the triple bottom line." In other cases, the dimensions are rendered as

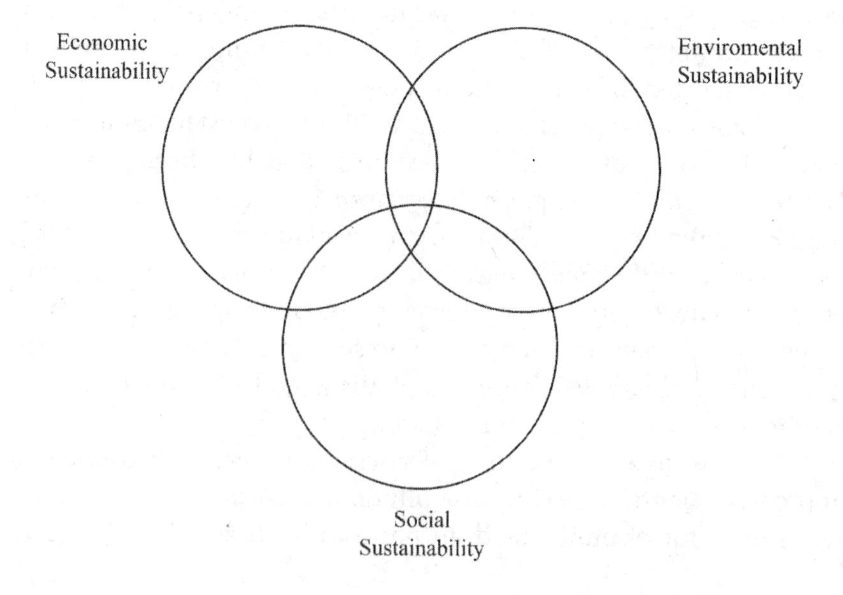

profits, planet, and people, leaving us with "the three Ps." Advocates of three-circle sustainability often cautioned against any attempt to go further in defining sustainability, though they were not opposed to discussion of indicators for sustainability. By the year 2005, it was clear that the broad idea of sustainability had been rescued from the years of acrimony. The National Academy of Science instituted a section in *PNAS* for sustainability science, and yet another spate of publications called for scientists to turn their research agenda toward more practical problem-oriented research that would move the world closer to the elusive goal of a sustainable society.[9]

THE TROUBLE WITH THREE-CIRCLE SUSTAINABILITY

We owe a great debt to the authors who proposed the simple-minded notion of overlapping dimensions, for in an era when disruption in global climate systems has joined the resource depletion issues that were noticed in the 1970s, sustainability is a more fundamental and important challenge than ever before. Viewed as a response to the unproductive debates of the 1990s, three-circle sustainability suggests a "just do it" approach in which researchers, policy makers, and the public at large simply get down to the business of promoting sustainability. Everyone has enough of an idea of what sustainability means, this "just do it" approach would suggest. Attempts at definition will simply slow things down by dragging people into theoretical and technical issues that really can be left to the specialists. While I do not doubt that there is a kernel of wisdom in the "just do it" approach to sustainability, there are reasons to be cautious.

Some of these reasons have been well articulated by others who are suspicious of the talk about profits and the so-called triple bottom line. They see the concession to an economic dimension as letting the camel's nose in under the tent. Soon, this line of thought has it, sustainability will be all about profit seeking, and there will be absolutely nothing in policies or standards for achieving sustainability that do anything to actually promote environmental goals. This is a line of criticism that takes off from earlier skepticism about sustainable development, seeing it simply as development by a different name.[10] More pointedly, this way of suggesting that even three-circle sustainability needs to be debated interprets putative concern with sustainability as a form of "greenwashing": making an environmentally indefensible practice appear to be consistent with environmental goals through advertising, public relations, and even outright deception.[11] In short, we need to take a critical attitude toward claims of sustainability in order to ensure that we are not fooled by unscrupulous interests having little true interest in sustainability.

There is, however, a more subtle way of developing this line of criticism that strikes deeply to the heart of three-circle sustainability. In listing three domains that are crucial to sustainability, three-circle sustainability comes very close to simply being a vague umbrella term that covers virtually any social goal. If the economic dimension is simply equated with business operators' need to make a profit, for example, there is an important sense in which the main point of economically motivated specifications for sustainable development proposed in the wake of the Brundtland report is lost entirely. As noted above, economists who advocated growth were *not* simply advocating perpetual profits! They were instead focused on the ability of the poorest people to sustain themselves and improve their quality of life. To understand the economic dimension of sustainability as the need for major international corporations to continue to make profits is a gross distortion of what post-Brundtland economists were advocating.

Similar things might be said about the environmental and social dimensions of three-circle sustainability. There are many environmental goals that may have little direct connection with the collapse of ecological systems. Preservation of charismatic megafauna, for example, is a cherished environmental goal, as is the more general protection of animal interests everywhere. Yet the connection between these goals and sustaining processes of soil fertility, the water cycle, or the general climate system is at least not obvious. The social dimension is especially problematic in this regard, as greater achievement of ideals of democracy, social justice, and improved well-being may in fact be contributing increased resource use in ways that are contrary to greater sustainability. If all that profit, planet, and people means is that we will now continue to pursue longstanding social goals under the new banner of sustainability, it is far from clear that sustainability means anything at all.

It is thus important to query the substantive content of each circle in three-circle sustainability to have any reasonable assurance that the programs and policies being adopted are genuinely progressive. This does not mean that we should return to the old days of "defining" sustainability. It is useful to think of sustainability in three overlapping ways, just as the three-circle model suggests. But it is even more useful if those dimensions are somewhat more carefully articulated than the triple bottom line model of economic, environmental, and social dimensions typically suggests. In fact, the three dimensions can be well characterized in terms of three ideals: resource sufficiency, functional integrity, and critical engagement.

BEYOND THREE-CIRCLE SUSTAINABILITY

What three-circle sustainability gets right is that there are multiple dimensions to sustainability. What it misses is that there are also multiple ways to *conceptualize* what sustainability is all about. An examination of the literature on sustainability suggests three distinct approaches. The first is *resource sufficiency*: A practice or system of practice is sustainable if and only if the resources needed to carry it out are foreseeably available. The second is *functional integrity*: A system of interacting parts is sustainable if and only if the elements that contribute to its operation and reproduction are resilient in response to external threats. The third is *critical engagement*: Social relations become unsustainable when practices of domination and resistance to domination threaten to silence the opposition and stifle positive processes of social change. Elucidating and specifying any of these three approaches within a given context becomes quite complex, but it is possible to clarify what each is about comparatively briefly. One of the main things that make the contrast between these approaches worth thinking about is the way that each sets up a somewhat different way to conceptualize the ethical issues arising from sustainability.

The idea behind resource sufficiency is comparatively straightforward. Most of the things humans do have the potential to consume resources. Food production utilizes sunlight, water, soil, seeds, and energy. Although each of these is a potentially renewable resource when measured over human time scales, only sunlight seems to be unaffected by the manner of production. Modern agricultural systems have become particularly dependent on the nonrenewable resource of fossil fuel. In addition, farming methods may deplete soils or fail to conserve genetic resources embodied in seeds. Other productive activities are even more dependent on the utilization of nonrenewable materials, including minerals and rare earths. To determine if any productive practice is sustainable, one needs to measure both the total amount of resource that is available and the rate at which it is depleted by the production practice. Measuring depletion of renewable resources will require one to understand more complex processes, such as the rate at which fish populations are reproduced or soil fertility is replenished. To understand the sustainability of consumption, one must know what production practices the consumption depends upon. This fairly simple conceptual outline of sustainability as resource sufficiency then needs to be augmented by recognizing that some consumption practices can be supported by several different production schemes, and that it is possible to shift both production and consumption from relatively scarce resources to relatively plentiful resources in some cases.

Economists who responded to the Brundtland report by arguing that economic growth is the basis of sustainability were making a sophisticated and egalitarian argument based on this possibility for shifting resource consumption. Their view was that once a society reaches a certain plateau with respect to its ability to meet the welfare needs of its population, the ability to shift from one pattern of resource use to another is dramatically increased through the aid of education, communication, and technology. However, poor countries have little flexibility in these matters. However, reaching this plateau will require an increase in the immediate short-term consumption of resources for at least two-thirds of the world's population. Thus from this perspective, one potential problem with the general idea of sustainable development is that simply putting a cap on consumption for poor nations has the effect of supporting the lifestyles of comparatively well-off people in the industrialized world, while condemning the rest of the world to perpetual misery and deprivation.[12]

This economic approach to sustainable development came to be known as "weak sustainability." It was challenged by a group of ecologists and philosophers who rejected the notion that a focus on economic growth would be in any way adequate for sustainability. In fact, they thought, economic growth was the main source of the problem. Bryan Norton characterized this view—strong sustainability—as a commitment to the preservation of ecosystems, natural capital, and a variety of goods that he insouciantly labled "stuff."[13] But what, exactly, is the debate between advocates of strong and weak sustainability about? There are at least three possibilities:

1. Advocates of strong sustainability reject the egalitarian ethical orientation of economists. They are quite willing to understand sustainability as requiring significant global inequality.
2. Advocates of strong sustainability accuse the economists of making mistakes in their calculation of the stock and flow of resources required for resource sufficiency.
3. Advocates of strong sustainability have a different conception of sustainability.

Advocates of weak sustainability are inclined to suspect that it is the first of these possibilities that is at work, so it is fairly important to get clear on this point. It is also worth noticing that both of the first two possibilities are consistent with the general conceptual framing of sustainability as resource sufficiency.

In fact, many of the ecologists who reacted against weak sustainability seem to be advocates of sustainability as functional integrity. They are focused on the integrity and resilience of ecosystems. The basic idea here is that ecosystems have a demonstrated tendency to "bounce back" from stressful events. A hurricane, a volcanic ash cloud, or an extreme climatic event can be viewed as a force or event that affects the ecosystem from the outside. An ash cloud, for example, has the potential to cause a widespread die-off in the flora and fauna of an ecosystem, yet within a matter of a few years, the effects of this die-off will not be noticeable and the mix and number of plants and animals will be much as it was before. This ability to bounce back or recover is what is referred to as resilience. However, some events are too extreme, and ecosystems cannot recover. So when we turn to human activities, it is possible to understand human use of or impact on an ecosystem as sustainable so long as the stress imposed by humans does not cross the tipping point and drive some element of the system into a state where it cannot recover. Overfishing of cod in the North Atlantic provides an excellent illustration of a human stress that exceeded the resilience of the fish population. However, the ecologists who study fisheries see humans as *part of* the ecosystem. They must regulate their fishing so as not to drive the fish population below the numbers needed to repopulate. Studies of traditional societies that depend on fishing have shown that they have a number of sociocultural ways to do this.[14] If the social mechanisms that regulate fishing break down and people fish too much, the total system—ecology plus society—becomes unsustainable.

People who approach sustainability through the lens of resource sufficiency tend to understand the ethical dimension in terms of the question "Enough for whom?" While this is clearly an important ethical issue, functional integrity opens up a broader domain of ethical issues that revolve around the way that we understand the borders of the relevant system and the mechanisms that regulate the system. The mere fact that this approach allows one to understand human beings and their social institutions as interactive components of the ecosystem is an achievement of ethical significance. It is arguably a step toward overcoming strong nature–society dualisms that have made it difficult to think about environmental values in ethical terms. What is more, it suggests that habits and institutionalized norms are important targets for ethical reflection and for conceptualizing the sustainability of human practices. Ethics, on this view, moves beyond managerial decision making and takes up the way that cultural *and* environmental feedback reinforces patterns of behavior that may play crucial functional roles from the systems perspective. Although it is entirely possible that one might utilize one and the same scientific models or data to calculate resource sufficiency as one uses to identify threats to functional integrity, the framing and conceptualization of these approaches place ethical or political conversations onto substantially different trajectories. For this reason alone, it may be important to debate sustainability.

CRITICAL ENGAGEMENT AND SOCIAL SUSTAINABILITY

One could argue that resource sufficiency and functional integrity capture most of what is discussed with respect to two of the three Ps, planet and profit. But do these ideas capture the full range of concerns that are brought up as aspects of social sustainability, the "people" dimension of the three Ps? Social sustainability is often interpreted in terms of advocacy for poor or marginalized human beings. Improvements in social sustainability may include better living or working conditions, more inclusion of economically disadvantaged or historically excluded people in decision making, and more effective realization of rights to health care, education, employment, and due process.[15] Thus, one way of asking the question about inclusion of the "people" P is to ask whether the indicators for social sustainability are included by analyses focusing on resource availability and functional integrity.

It is possible to answer this question in the affirmative. As noted already, some of the economists who developed the weak sustainability approach did so because they believed that ensuring economic growth in poor countries would contribute to improvement in the quality of life for poor and historically marginalized groups. However, the neoclassical economic theory of welfare economics has a number of notorious weaknesses that make it relatively easy to overlook inequities in the way that gains from economic growth are distributed. Thus a more promising way to think about incorporating social sustainability into resource sufficiency is to utilize the capabilities approach to human welfare advocated by Amartya Sen.[16] Sen's approach is both more sensitive to the way that social changes affect the substantive freedom of individuals and more attentive to the impact of change on an individual's actual ability to exercise rights and freedoms, including rights of participation and due process. Similarly, the need to replenish social capital might provide a way to incorporate the concerns of social sustainability in an approach that stresses functional integrity. Like the natural capital associated with ecosystem services, social capital is conceptualized as forms of trust, social cohesion, and institutional solidarity that make social cohesion possible.[17] An approach to sustainability that was sufficiently attentive to capabilities and to social capital might be said to address many if not all of the key indicators typically associated with social sustainability.[18]

Other authors have argued that goals associated with social justice are better understood when they are not buried in technical models of resource sufficiency or functional integrity. Concern for inclusion of poor or marginalized groups can be articulated as an ethical or political priority wholly apart from our goals of achieving sustainability, so it is possible that one would take the position that these goals are better understood and appreciated when they are seen as having nothing whatsoever to do with sustainability.[19] Still others have argued that sustainability should be understood as a generic term that names a broad social movement, rather than as a concept that expresses a particular goal or content. On this view, it is important to include aspects of social justice under the banner of sustainability simply because it is only meet and right to do so.[20] There are thus a number of perspectives that might be taken on the relationship between the indicators of social sustainability and various conceptual or technical approaches to sustainability. The multiplicity of perspectives is all the more reason to think that sustainability is worth debating.

However, recognizing this multiplicity of perspectives does not appear to require a third conceptually distinct way of approaching sustainability. Yet the forms of protest and resistance that surround actions to promote *any* kind of sustainability might be interpreted as pointing toward a unique approach to sustainability, one that is not very adequately reflected in the ideas of resource sufficiency or functional integrity. This approach might start by noting that oppression has been met by resistance during every epoch of human history. It might go on to argue that the give and take between oppressor and the oppressed has been a driver for social change and progress throughout history. However, many theorists of social change have argued that when the modern bureaucratic state and the modern multinational corporation become equipped with modern communication and advertising technologies, this healthy dialectic becomes subverted, driving resistance underground and making participation in the political process ineffective.[21] This line of thinking might be seen as relevant to sustainability to the extent that closing off opportunities for critical interaction and amelioration of the tensions that arise in normal processes of domination and resistance become a threat to the stability of societies, or bring processes of adjudicating injustices to a halt. Some theorists have used the idea of social sustainability in just this sense.[22] Others have argued that attempts to roll resistance movements into sustainability efforts such as eco-labels and product standards only wind up subverting the goals of resistance,[23] so there is plenty to debate here as well. But what should we call this general line of thinking on sustainability? Is this social reflexivity? This kind of terminology is indeed used in contemporary social science,[24] but I have no wish to detain readers with even *more* introductions to new words and difficult theoretical traditions. With homage to this tradition's debt to critical theory, I call it "critical engagement," and leave the idea at that.

CONCLUSION

We may not need to *define* sustainability in order to interact in a useful way in pursuing sustainability. Yet scientists and policy makers in particular are apt to find themselves a little surprised and deeply confused by the plethora of meanings and approaches to sustainability if they fail to participate at all in the debates over what sustainability means. When Thomas Jefferson and John Adams found themselves serving under President George Washington, both men came to appreciate that they had rather different understandings of the word "democracy," a word whose meaning they had only recently been willing to take up arms in order to defend. They did not "define" democracy, though they did debate it. Their debate has served us well. It is likely that generations two centuries hence will be thankful if we do the same for sustainability.

9.

BIOTECHNOLOGY AND THE PURSUIT OF FOOD SECURITY

David Castle

INTRODUCTION

The world's population is growing; so too is the number of people who lack food. Recent estimates place the number of hungry between 963 million[1] and 1.2 billion[2], troubling numbers even if they were static, which of course they are not. Increasing populations, particularly in regions with historically poor agricultural productivity, combined with global food price volatility, is expected to make the situation worse and much more acute in some regions. Multiple solutions are being considered, including better distribution of food, utilization of marginal land, and techniques for improving yields and food quality simultaneously. This chapter focuses on the contribution that science and technology can make to improving food security in its different aspects. The problem of food security can be defined in basic terms of caloric deficit, but other aspects of food security, such as the reliability of supply and the cultural appropriateness of available food, must be considered. Hunger is a multi-faceted problem, and like all complicated problems nuanced by local factors, its causes are likewise complicated. These include population growth, perhaps the most obvious and direct driver of food insecurity, and other threats, such as political instability, competing resource allocation, water shortages, and transitory and enduring effects of climate change. Hunger is also caused by persistent food access inequality between nations, within them, and even within households. Income disparities, class structures, cultural divides, and traditional views about the roles of women and children reinforce imperfect distribution and access to food.

There are reasons to be doubtful that population growth can be curbed and distribution and access so vastly improved that global food security will be a reality within the next few generations. Critics are equally pessimistic of boosting agricultural productivity, but there is a difference of fact that opinion cannot erode. If one had an accountant's ledger for food security, the red ink on it would be due to population growth, particularly in cities, worsening income disparities worldwide, and mounting competition for water, land, and sea. The main reason there is any black ink on the ledger is due to the increases in agricultural productivity over the last one hundred and fifty years. The shortlist of factors contributing to increased productivity includes mechanization, inorganic fertilizers, irrigation, high-yielding hybrid varieties, improved crop rotation and adoption of minimum and zero tillage, and precision farming aided by global positioning systems. Looking back over the last fifty years, there have been tremendous productivity increases in what are now called the terrestrial Green and aquatic Blue Revolutions.

Productivity gains would not have occurred without science, technology, and engineering, and genomics and genetics and associated biotechnology will play an increasing role. Despite the evidence of gains, the paths to sustainable productivity gains and the role of genomics and biotechnology are widely and hotly disputed. The dispute often arises from disagreements about how food security should be defined and measured. Universally accepted metrics for agricultural productivity are few in number, making it difficult to prioritize and to reach consensus about definitions of success. The absence of clear definitions and metrics for food security and productivity do not, however, diminish advocacy for different solutions. There is strong and often needlessly polarizing advocacy for organic versus conventional-industrial agriculture and the role of genomics and biotechnology enhanced production. To complicate matters further, empirical and ethical claims are often run together in arguments, leading to situations in which sweeping claims about the right or wrong way to proceed are based on thick ideology and thin evidence. What role should genomics and biotechnology play in achieving greater productivity in pursuit of food security, and how should paths toward sustainable productivity gains be set?

THE PROBLEM: FOOD SECURITY; FOOD INSECURITY

Food security has been variously defined in its forty years of widespread use. In a recent review article, Per Pinstrup-Andersen, winner of the 2001 World Food Prize for his contributions to global food policy, observes that food security once linked food with national security or sovereignty.[3] At first, the measure of food security was the total food calories available to a population divided by the number of people in the population. Food security has since had another three elements added. First, supply of calories is not the same as access; within a country, or within households, there can be profoundly unequal access to food. Second, nutritional value and food preferences need to be considered; secure food has the dimension of quality and quantity. Third, although caloric availability is an important measure of food security, when used as the sole benchmark for food security it can be misleading. Today, an estimated one billion people lack adequate daily calories, but if iron deficiency is the benchmark of food security, the number doubles.[4]

The distribution of global food security follows easily predicted patterns. Food security tends to be a persistent problem in lower and middle-income countries (LMICs) where transient shortages occur frequently against a background of inequitable distribution. Based on the 2006 estimate of the Food and Agriculture Organization of the United Nations (FAO), which is out of date but useful for proportions, 845 million were undernourished, of which 820 million were in LMICs, a number which has changed little since the 1990s.[5] While the FAO in 2006 predicted that fewer (582 million) will lack food security in developing countries by 2015, reliable predictions are elusive because of the number of factors conspiring against food security. The remainder of food insecure people reside in thirteen transitional economies, such as Belarus, and developed countries, such as the United States, where as many as 13 million children lack food security.[6] Global repercussions of financial crises will undoubtedly leave fewer without adequate and nutritious food in many cities in industrialized countries, suggesting that food security may become a more apparent phenomenon in unfamiliar contexts.

Food *insecurity* is the problem, and it has many dire consequences for individuals and countries. For individual health, both persistent and transitory shortfalls in macro and micronutrients cause development problems, risks to mothers and fetuses, risk of chronic disease, susceptibility to infection, blindness, stunted growth, cognitive impairment, and aggressive behavior—and these are the fates for those who do not simply waste away and die. In striking contrast, for a segment of the global population, food insecurity involves too much simple macronutrient availability but poor overall nutrition and the metabolic syndrome—hypertension, obesity, diabetes, and dyslipidemia—is on the rise. Apart from the effects on individuals, the social impacts of food insecurity include the spread of communicable diseases, theft, disputes about access to food and to land and livestock, harms specifically to women and children, and social unrest and violence. Food insecurity cripples economic growth, and can trigger corruption. Furthermore, food insecurity can cause civil war, can contribute to conflict between nations, and can be exacerbated by, or play of off, underlying cultural or existing geopolitical tensions.[7]

The factors that are associated with food security for a nation or region include the level of economic development, type of political system, extent of literacy and numeracy, climate and geography, and history. Yet these factors cannot be simply combined as if they were ingredients in a recipe for food security. The dynamics between factors that cause food secure nations and regions to suddenly become food insecure are not easily predicted, and some food secure nations and regions enjoy relative food security without having a full slate of positive socioeconomic attributes. Because food insecurity is characterized by the FAO as shortfalls in macro and micronutrients, as well as the lack of culturally appropriate foods that would meet nontrivial social or religious preferences,[8] it can be difficult to align the socioeconomic factors that contribute to food security with FAO criteria.

Pinstrup-Andersen questions whether the FAO definition "can be used to guide policies and programs or whether there is a need to disaggregate the concept into different kinds of food insecurity depending on the nature and severity of the problem and the type of solution required."[9] Because efforts to fully characterize the problem of food security tend to expand the definition of food security from a basic conception of caloric deficiency, the risk is that different yardsticks and thresholds will create an account of food security that is as complex as the phenomenon itself. This is a legitimate concern because it highlights the fact that food security can be sustained, or food insecurity can persist, in different places and at different times for a whole host of reasons. The implications and meanings for individuals, households, societies, and nations are similarly varied.

POTENTIAL SOLUTIONS FOR FOOD INSECURITY

One of the *Millennium Development Goals* is to halve the number of those facing "extreme poverty and hunger" by 2015.[10] This laudable goal will not be achieved, except perhaps by sleight of definition and metrics. The reality is that at the time of writing, there remains a potent mix of increasing population, distributional inequity, and mixed prospects for increased agricultural productivity. These are the known factors; the unknown factors include the potentially devastating effects globally or regionally of climate change,[11] the destabilization of societies caused by increasing urbanization,[12] and the effects of global water cycle changes.[13] Keeping with the known factors, population, distribution, and productivity, it becomes clear that science and technology will impinge directly on productivity but will not make much of an impact on distribution and population.

With respect to populations, the obvious dread factor is the global increase in human populations. There are now more people alive than have existed throughout history; this is a sobering thought when the human population is projected to be over 9 billion by 2050. It is difficult to imagine what the world will be like with half as many people again as there are now, especially since the global trend toward increased urbanization alters agricultural practices as smallholder farms decrease in number. If people are not food insecure in rural areas, they will become food insecure in cities with more grievous social consequences. Not since the Club of Rome has there been sustained and serious public discourse around population control. Indeed, the coercive measures in India in the 1970s and the ongoing demographic problems induced by China's one-child policy have put a damper on the topic. Perhaps there is some long-term room for optimism, since as the population grows in absolute numbers the actual growth *rate* has been declining worldwide since 1962.[14] Some take this as evidence that unconstrained Malthusian growth is actually limited by social factors. As populations gain food security the death rate falls and the population increases, but then birth rates decline, particularly if family income rises. Because death rates fall faster than birth rates, there is a temporary population boom. The theory of "demographic transition" links affluence and a declining population, and is used to explain the aging and declining populations in industrialized countries,[15] although there is recent evidence that the transition can be reversed.[16]

Population growth puts obvious pressures on food security, and the rise in the number of people surely makes it unlikely that the hunger and poverty Millennium Development Goals will be achieved. Even if the rate of population *growth* declines, pressure on food security also comes from the dynamics of the demographic transition. For example, people subsisting in food insecure regions get fewer than twenty grams of protein a day, most of it from plant sources; people in developing countries usually get more than twenty grams of plant and animal protein; and people in industrialized countries easily get more than ninety grams of animal protein each day.[17] Furthermore, the boom caused by the more rapidly declining death rate compared with the birth rate causes a minimum of two generations' increase in population needing food security.

With respect to the distribution of food, it is often remarked that enough food is currently produced worldwide and, were it not for unequal distribution, all would have enough to eat. This is true in terms of total caloric output, which the FAO in the late 1990s estimated at 2,620 calories per person.[18] Total caloric output is not a measure of nutritional balance when averaged for all people, and further, the major increase in calories produced in the five decades since the 1960s—approximately 25 percent—is still outpaced by population growth in the same period—approximately 85 percent. The population is expected to increase by another half in the next forty years, but forecasts of increased agricultural productivity do not keep pace.

These are important considerations to bear in mind when considering the distribution problem. In the late 1980s and early 1990s the study of social determinants of the food security problem often focused on the distribution problem. Amartya Sen convinced many, on the basis of studies in India and some sub-Saharan African countries, that if one accepts there is enough food to go around food security could be viewed as a demand-side problem. The problem is that people in LMICs lack purchasing power, and without the "entitlements" held by others, they were unable to access food. Consequently, the distribution problem could only be solved through poverty alleviation, thus breaking a vicious circle of food insecurity stifling economic development, which, in turn, makes achieving food security more difficult.

Distribution is often related to food aid where rich and food secure nations can donate food or use their wealth to purchase and redistribute food. Food aid makes sense on humanitarian grounds in situations of acute food insecurity, for example if crops utterly fail or in cases of war- or natural-disaster-torn societies. Growing criticism about food aid to address persistent food shortages is growing, however. Dambisi Moyo argues that aid reinforces the poverty traps many nations find themselves in by perpetuating dependence on foreign resources rather than developing domestic capacity.[19] This would be true for loans as well as direct material aid. The notion that food security is connected with food sovereignty reinforces the point that attempts to distribute food aid must overcome issues associated with sovereignty, a fact that becomes problematic when other factors such as sanctions, debt, or resource depletion apply.[20]

In light of the problems associated with food distribution, developing domestic capacity for self-sufficiency and for trade is paramount. Compared to the cost of providing direct food aid annually, providing local assistance to farmers is far less expensive. One estimate of the cost to buy, ship, and distribute a tonne of US maize in Africa is US$812; to buy it locally in Africa and redistribute it is US$320; and to give farmers the fertilizer, seed, and support to grow an extra tonne of maize is $135.[21] This idea of direct subsidy, often through voucher programs similar to a microcredit system, is gaining momentum in Eastern Africa, where food shortages are common and severe. A recent success story in Malawi illustrates the point. After some of the worst harvests in a decade, the Malawi government invested in, and sought support from, the United Nations, for an improved seed and fertilizer voucher program. Planting in 2006 coincided with good rains, and significant yields were achieved. The program was repeated in 2007, leading to a bumper crop, some of which was exported to Zimbabwe.[22] A similar program has been developed in Kenya, where subsidies were given with decreasing amounts over the course of several years as farmers gained autonomy. The program required 10 percent of yields to be donated to schools, and farmers were required to repay the subsidies from profits gained through improved yields. The rate of repayment was 95 percent, and it has created a revolving fund that allows for future capacity building.[23]

The driving force behind the Malawi and Kenya examples is the need to increase local agricultural productivity to break the poverty trap that exacerbates food insecurity.[24] The main elements in crop production systems are improved seed and the appropriate use of inorganic fertilizers to boost yields. Farmers in these contexts can neither wait for a demographic transition nor rely on foreign aid indefinitely. For many, the problem is simply not having a coordinated, systematic, and integrated farming system in their country, and instead relying on very poor quality seed, planted without irrigation (rainfed only), no fertilizer or weak organic fertilizers, in leached, acidic, and often highly aluminum toxic soils. Food security, therefore, needs to be linked to the question of how science and technology can be used to address this suite of problems.

Because slowing population growth is unlikely any time soon, and redistribution of food is at best a partial and challenging solution, productivity gains in agriculture must be sought. Jennifer Thompson, a scientist heavily involved in the food security of Africa, puts the point forcefully:

> Many people say enough food is produced in the world to feed everyone. However, people have been saying this for decades and we still have shortages. One of the problems is distribution; how do we get food to the people in need? Certainly, we should stop wars, eliminate corruption so that food gets to the right people, build roads and rails to transport food, but how long will that take?[25]

Thompson's view is that improvements in infrastructure and social coordination, as well as better distribution are all needed, but without a commitment to improve yields, current and future food security challenges will not be met. When one compares the improvements that have been made in agriculture in industrialized countries compared with Africa, the gap is obvious. Take for example sweet potato, a staple in Africa, where the average yield is 6 t ha^{-1} compared with the global average of 14 t ha^{-1}.[26]

Like Florence Wambugu, founder of African Harvest, who had attempted a genetically modified (GM) cassava project, Thompson believes that "GM crops that give increased yields are just one of the ways in which we can tackle the problem."[27] Thompson identifies four urgently needed improvements in crops in Africa that would bolster food security: insect resistant African Maize varieties expressing *Bacillus thuringiensis* (*Bt*) *cry* genes that are lethal to corn borers to improve yields and reduce mycotoxin loads; resistance to the maize streak virus and African cassava mosaic virus; maize resistance to the parasitic weed, *Striga*; and drought tolerance.[28] The strategy adopted by Thompson, Wambugu, and others is to realize that so long as rainfed crops are vulnerable to drought and, in turn, virus and insect attack, the cycle must be broken, and dramatically. Having met and discussed these issues with Wambugu and Thompson, the author is convinced that both biotechnology advocates would use any available strategy for crop improvement, genetically modified or otherwise, but genetic modification holds the most potential for dramatic differences—were the technology developed with applications for Africa and other developing countries specifically in mind.

The preceding statement is highly controversial; many disagree that commodity crops that are modified for various traits are more productive and more sustainable. There is a vast literature on this subject, most of which is plagued by irreconcilably different methods when methods are cited, and few studies have been replicated. In addition, many studies are not independent of agribusiness company data, which is not to accuse companies of being misleading, but to say that some claims have not or cannot be verified. There are also issues related to technology development and adoption, such as farmer's privilege, intellectual property rights, biodiversity effects on target and nontarget species, weediness, arms races between pests and active ingredients, choice of traits, and so on, which deserve and have received careful scrutiny.

The controversy surrounding any statement regarding agricultural biotechnology, particularly genetic modification, can divide opinion into pro- and antiscience, pro- and anti–big business, and so forth. Increasingly, the problem regarding agricultural biotechnology is not the technology itself, but the inability to overcome impasses in communication regarding the risks and benefits of biotechnology. This point is particularly well illustrated in Robert Paarlberg's *Starved for Science: How Biotechnology Is Being Kept out of*

Africa.[29] Paarlberg argues that there is a good case for using the technology of genetic modification to bring about rapid improvements in African crops. Yet the technology is kept out of Africa by advocacy groups in industrialized countries that have built their foundation of opposition to genetic modification where food security is less an issue than it is in Africa. This is a troubling phenomenon; yet Paarlberg's insightful account explains how well-intentioned people like genetic-modification opponent Vandana Shiva can appreciate the global scale of the food-security crises,[30] yet fail to engage adequately with the Jennifer Thompsons and Florence Wambugus of the world.

Considering the definitions of food security considered above, and the Gordian Knot of population, distribution, and productivity, two generalizations about biotechnology and food security can be offered. The first is that causes of, and solutions to, food-security problems must always be contextualized. The second is that agriculture, all of which must be considered technological in contrast to nomadic gathering, offers many diverse solutions to contextualized problems. The former observation will be discussed below. With respect to the contextualization of problems and finding solutions from a wide array of technologies, displaced ideology often constrains options or interferes with implementation. One dimension of this problem is that there remains a serious gap in knowledge about global agricultural productivity—how much is produced, where, and by whom, and what the bottlenecks are to greater productivity.[31] Despite efforts by the FAO, the Organization for Cooperation and Development, and the fifteen centers of the Consultative Group on International Agricultural Research (CGIAR), a systematic, networked initiative has yet to be launched.

Another dimension is being able to characterize problems such that they lend themselves to solutions. Klaus Amman, in this respect, has written two compelling articles on a new paradigm for agriculture that embraces whatever solution will produce sustainable results.[32] Amman is not preoccupied with an ideological rapprochement between pro- and antibiotechnology camps, preferring instead to encourage the adoption of methods and technologies for which there is evidence of efficacy. We need to bypass the rhetoric of entrenched positions described by Paarlberg because, as Amman says, "Human beings should be part of any risk assessment in technology: this is a request with enormous ethical implications."[33] This should not be confused with uncritical technology endorsement; hyping technology widens the gulf for failure. Amman's view is that agricultural management must scope over organic- and genomics- and biotechnology-based practices if it is to take into consideration the biodiversity effects of agro-ecology. Continuing to pit "naturalness" and "intrinsic value" arguments against the mantra of the continuity of transgenics from past breeding practices recapitulates an old debate, one which threatens to hamstring discussions about future—and possibly nontransgenic—agriculture.

Regarding an Evergreen Revolution, Amman advocates for a synthesis of tools for specific jobs, including global-information-system-enabled precision farming using biotechnology, improved crop rotation and soil management, and measures to improve on- and off-farm biodiversity. He says, "We can only achieve the goals of sustainability successfully if we are ready for progress, changing the world and the course of evolution, thus producing the best outcome for biodiversity, humanity and our planet."[34] With respect to appropriate biotechnologies for an Evergreen Revolution, one could include Gressel's idea of rescuing "orphan crops" and pushing them through a "genetic glass ceiling" to achieve greater crop biodiversity,[35] or using "zinc finger" techniques to place genes of interest in precise positions in a plant's genome.

Individual technologies will have to be tried and tested, but they constitute part of the path toward what the British Royal Society calls "sustainable intensification of agriculture." Among the many recommendations made by the Millennium Project Task Force for Science, Technology, and Innovation, it was suggested that platform-scale technology development is crucial to achieving the Millennium Development Goals.[36] This same perspective has been echoed recently in the World Bank Report 2010. The third chapter of *Development and Climate Change*,[37] "Managing Land and Water to Feed Nine Billion People and Protect Natural Systems," describes predicted decreases in yields, particularly throughout the tropics and subtropics, by 2050 as the result of climate change. Temperate regions may see increased productivity with warming, all other factors being equal, but the implication from the report is that there will be instability in agricultural regions, making potential gains and losses difficult to predict. The risk is that many biomes not currently under agriculture will be converted, potentially exacerbating the effects of climate change through deforestation. The report calls for biotechnology development in a vein similar to that suggested by Amman and Thompson; that is, on a "case-by-case basis, comparing the potential risks with alternative technologies and taking into account the specific trait and the agroecological context for using it." The path forward for food security includes biotechnology:

> Biotechnology could provide a transformational approach to addressing the tradeoffs between land and water stress and agricultural productivity, because it could improve crop productivity, increase crop adaptation to climatic stresses such as drought and heat, mitigate greenhouse gas emissions, reduce pesticide and herbicide applications, and modify plants for better biofuel feedstocks.

CONCLUSION

The outlook for a hot, hungry, overcrowded, and inequitable planet, now, and in 2050, is grim. If one in six are food insecure now, what will the numerator be when the denominator is 9.5 billion? And can anyone seriously claim to predict how climate change and urbanization will shape food insecurity? The lessons learned in the last century and a half, since mechanized agriculture relieved people of toil, is that while some prosper, others do not for reasons as complex as the phenomenon of food insecurity. Curbing population growth and tackling poverty remain crucial, yet at the same time something must be done to improve agricultural productivity—terrestrial and aquatic. Systematic assessment of the impact of genomics and biotechnology, coupled with other scientific approaches to soil, fertilizers, and irrigation comprise the elements of an evidence-based path toward food security. There are other issues to address, of course, such as the problems associated with social coordination related to effective knowledge management,[38] the need to recruit properly to the challenge,[39] and the need to find strategies for coping with constraints imposed by intellectual property.[40] These are the necessary supports to create sustainable intensification of agriculture for enduring food security.

Section 2.

CLIMATE CHANGE

10.

ETHICALLY DEALING WITH CLIMATE CHANGE

Comparing the Maldives, China, and the United States

Bill McKibben

Climate change is the first truly global problem the world has ever faced—but it faces it in a world split up into different nations with different histories and different destinies. So it's both a fascinating and a fateful question to try and figure out the duties those nations owe each other—the ways we answer this question, which is both an ethical and practical query, will likely decide how the twenty-first century comes out, and perhaps a great many centuries that come afterward. In fact, you could say it's a test of how big both our brains and our hearts really are. And so far the answer is not at all clear.

Let's set the ground rules first. Given our distance from the sun, our atmosphere has to do the job of maintaining a livable temperature, something that, say, Venus and Mars have failed to accomplish. Trace gases in the atmosphere, especially carbon dioxide, play an important role by trapping some of the sun's heat that would otherwise radiate back out to space. Too little and we'd be very cold, but too much and the sweet spot we've inhabited climatically for the ten thousand years of human civilization, a period scientists call the Holocene, will start to heat. In fact, it already has. For two hundred years human beings have been burning coal and gas and oil, and hence the atmospheric concentration of CO_2 has grown from 275 parts per million to almost 400 parts per million. This is very bad news because in 2008 a team of NASA scientists declared that any value for carbon in the atmosphere greater than 350 parts per million is not compatible with "the planet on which civilization developed and to which life on earth is adapted." Because we've passed that threshold, the Arctic (and indeed every other frozen thing on the planet) is melting; the sea is turning steadily more acid; and the atmosphere is much moister, meaning a great increase in downpour and flood. In order to return slowly to the 350 level we'd have to work extremely quickly to move the world off coal, oil, and gas. And that, of course, is very tough to do: cheap fossil fuel is the heart of the modern economy. Moving quickly to get off it would require money and government focus. It would be, to use Al Gore's word, "inconvenient."

So let's look at the circumstances of different nations, and how they are coping with that inconvenience.

First the United States. As the world's most affluent large society, and the place where fossil fuel use took its most extravagant form, it is no wonder that America has produced the largest share of the CO_2 now heating the planet. Even now, with the rise of other powers in the last decade, the 4 percent of the planet that is American produces more than a fifth of the earth's carbon. The average American produces ten times more than the average citizen of the developing world.

So far our response to the need for energy reform has been very limited. After dismantling early renewable efforts (like the solar panels on the White House roof that came down during the Reagan administration), the US government has served mainly to obstruct international action on climate. It was the only major industrial power that refused to honor the Kyoto accords negotiated in the late 1990s that would have begun, albeit slowly, the reduction of greenhouse gases. During the Bush administration even government scientists who acknowledged global warming were harassed and censored. The Obama administration has recognized climate change as a critical problem, but its pledges—a 4 percent reduction in US emissions from 1990 levels by 2020—fall far short of what other industrialized countries have undertaken.

Why has America not acted faster? One reason is the power of vested interest: some of America's most profitable companies are in the fossil fuel sector, and they wield considerable power. For the past several years, for instance, ExxonMobil has made more money than any company in the history of money. Another reason: Americans have become used to low energy prices, and rebel any time they begin to rise, so politicians have been reluctant to take steps, like a gas tax, that would reduce consumption. A third reason, perhaps, is that polling data shows Americans seem to believe that any effects of climate change will be slow to appear, though scientists have demonstrated that this is not the case.

Our second nation, China, can see with great clarity the potential effects of climate change—the Yellow and Yangtze rivers, on which it depends, come from the dwindling icepack of the Himalayas, while the industrially rich Pearl River Delta is only meters above a rising sea.

In most respects, the Chinese cannot yet be held liable for much damage. Though in 2008 they passed the United States as the largest single emitter of carbon, their emissions are relatively recent, since large-scale industrialization has happened only over the last fifteen years. Given that carbon molecules have a residence time in the atmosphere of a century or more, Chinese emissions still make up only a small fraction of the total. Equally important for any ethical considerations, there are far more Chinese than Americans. Even as they've passed the United States in total emissions, the average Chinese still uses barely a quarter the energy. (If you think this shouldn't matter, consider a thought experiment. Say China broke itself into four countries, each with the same population as the United States, but each with only a quarter of its emissions. Would that solve the climate problem?)

Moreover, unlike the United States, the Chinese have been using their coal-fired power to pull great numbers of people out of poverty. Most of the world's decrease in poverty in recent years comes from Chinese leaving the hard rural life for a somewhat more cushioned urban existence. There remain perhaps 350 million Chinese—more than the total population of the United States—who demographers believe would like to make that same move in the next few decades.

Caught between the trouble of global warming and the demands of its economy for more growth to accommodate that migration, the Chinese have done two things. One is burn ever more coal—at some points in the last decade they were opening a new power station weekly. But they have also invested far more money in renewable energy than other countries, and with very concrete results: China may have half the world's installed renewable capacity, mostly because 60 million households (250 million people) use hot water heated in solar thermal panels on their roofs. In recent years China has taken the lead as the fastest installer of wind turbines, the biggest manufacturer of photovoltaic panels, and so on. This race between what you might call green and black energy is hard to handicap—both seem to be surging.

Globally, China has rejected calls for capping its emissions, arguing that it has a right to emit as much per capita as Western nations. But informally it has set a goal of peaking carbon emissions by 2030. In the meantime, it has promised to reduce its energy intensity—the amount of carbon produced per unit of output—by 40–45 percent over the next decade.

A few things are made clear by this summary.

One is that the United States has no moral ground on which to stand in demanding that China reduce emissions at the same pace as developed nations. That argument amounts to nothing more than "we filled the atmosphere so you find some other way to develop." That seems ethically problematic to say the least: surely a Chinese person has no less right to the atmosphere than any of us. And at this point, if burning more coal is rapidly reducing poverty, they may have a better claim to *more.*

Another is that, given rapid Chinese growth, there's no way to address the problem of climate change without ensuring that their emissions peak as quickly as possible. Practically, China will wreck both its own future and the planet's should it continue to grow its emissions

The "normal" way you might solve this dilemma is for the nations that caused most of the problem to compensate the ones that suffer as a result—for instance, the United States might transfer technology and resources to China to make up for the fact that they would need to use more expensive technologies than coal. This is happening in a very limited way—but politically it's extremely difficult. Imagine running for Congress on the platform that we should send money to China.

As a result of this ethical, practical, and political tangle, very little is in fact happening. The Copenhagen climate conference in December of 2009 foundered on these shoals, and instead of a vigorous global agreement, China and the United States led the negotiation of a very limited accord. If you take the promises that the Chinese and Americans and other nations make under that accord, computer modeling indicates that the planet will have a CO_2 concentration nearing 700 ppm by century's end. In other words, if the planet is not precisely hell, it will be a similar temperature.

So let's consider one more nation, one of the smallest on earth. The Maldive Islands in the Indian Ocean are paradisical: it's an archipelago of 1,200 small atolls, most barely bigger than a football field, each ringed by its own reef. White sand beaches, palm trees, warm trade winds.

But it has several problems. A warming and acidifying sea has combined to kill much of the coral that provides the islands' fishery, draws many of its tourists, and protects the beaches from the pounding of waves. Worse, the highest point in the archipelago is only a couple of meters above sea level. Predicted rises in the ocean from melting ice packs mean that the 300,000 residents may well be refugees by later in the century.

Though poor, the nation is preparing for that possibility, laying aside some money each year to buy a new homeland, though no one knows where that might be.

More to the point, the Maldives has also undertaken a path-breaking commitment to clean energy: its president, Mohammed Nasheed, pledged in 2009 that the nation would be completely carbon-neutral by 2020. Already contracts for wind turbines have been drawn up, and experiments are underway with turning sea-weed into fuel for the boats and motorcycles that are the nation's main transportation.

Perhaps most importantly, the nation has engaged politically more than any country on earth. In October of 2009, for instance, Nasheed trained his entire cabinet in scuba diving so that they could hold an underwater cabinet meeting that passed a symbolic resolution demanding that the United Nations set the 350 ppm threshold as its target. (The event, needless to say, drew enormous coverage). Nasheed was the first foreign leader to arrive at the Copenhagen conference, and he pressed during and after for more action. Even at the personal level, he insists on riding a bicycle to his presidential office.

In other words, the Maldives has chosen to do everything that it can to try and meet the crisis, while an observer might be forgiven for thinking that the United States has chosen to do as little as it can get away with and China has split the difference. In ethical terms, I think, this small nation newly emerged from autocratic rule is setting the pace for the big and powerful. In so doing, it may serve as an example for how all of us should act in a time of crisis: with everything that we have at our disposal, and with as much thought to the future as the present. The last time I was in the Maldives, a ten-year-old girl, at the end of my speech, asked me if I thought she'd still be able to live in her home when she was a mother. I was able only to say that I would do all that I could to try and make sure the answer was yes.

11.

SCIENCE, ETHICS, AND TECHNOLOGY AND THE CHALLENGE OF GLOBAL WARMING

Albert Borgmann

Science, ethics, and technology all come under the heading of human enterprises. Science gives us the most general and rigorous insights into the structure of reality. Ethics gives us authoritative guidance for our conduct. Technology is the application of scientific insight in the service of an ethically commendable goal, "the relief of the human estate."[1] But evidently something went wrong in the interplay of these three projects. They have failed to meet the challenge of global warming. What's so remarkable about this failure is the apparent blamelessness of the scientific and ethical theories. They have done their work well in analyzing and evaluating global warming, and yet we can be anything but confident that science and ethics will make people do what needs to be done to address the current injuries and catastrophic prospects of climate change.

What's missing, I want to suggest, is a third theory in addition to scientific theory and ethical theory—a theory of technology. Whether such a theory will help us to meet the challenge of global warming, who is to say? It's in any event, a philosophical challenge to explain the mindless inertia of this country, its steady and apparently inexorable movement toward disastrous global warming and injustice. What a theory of technology can contribute remains to be seen in what follows. But that there is something that's poorly understood about the culture of technology should be readily evident.

To begin with the global dimensions of climate change, it's helpful to recall the global conquest of the human species. There have been at least three phases in that adventure. The first phase came to a conclusion roughly fifteen thousand years ago when humans had invaded and settled the last continent, what was then and still is now the new world, America. But although humans had by then covered essentially the entire globe, their awareness of the earth was local. Each tribe thought of its place on the globe as the center of the universe and of the rest of the globe as an indistinct periphery.

Humans for the first time became conscious of the globe in its entirety through the journeys of discovery Europeans undertook in the sixteenth century. Sailors, soldiers, and missionaries knitted the several local conceptions of the world into a global awareness. This was the second phase of globalization, soon to be intensified by the Industrial Revolution.

Global awareness is one thing, global integration another. We are now in the midst of the third kind of globalization where the last indigenous and traditional local cultures are being absorbed and transformed by the global culture of technology. There is a logic to the progress of globalization. It has moved from slow adaptations to instant innovations, from stable cultural diversity to accelerating technological uniformity, and, most important, from innocent ignorance to arrogant thoughtlessness—an unsuspected effect of industrial and technological globalization has been disastrous global warming. It has radically changed the material and moral atmosphere. Although the spatial scope of globalization has been finite ever since humans have had a global comprehension of the earth, the intensive scope of global transformation had seemed endlessly open prior to climate change. There was apparently no limit to human energy. But then the doors to intensive progress quietly shut, and now the use of the kind of energy that's been most familiar and effective for us is making life within this confinement more and more perilous. We're facing an inescapable danger while deprived of our customary energy. And we are caught not only in a physical predicament, but in a moral one as well. Whatever physical action I take affects everyone else on the globe, and while my individual action is globally inconsequential, when everyone in this country reasons that way, the consequences for other countries are enormous, and if all the prosperous peoples act in self-centered ways, the effects on the poor will be catastrophic.

Global warming and global justice have become indissolubly linked, and this linkage has given global warming the kind of daunting urgency that is not unlike the specter of a global nuclear holocaust that has haunted the second half of the twentieth century. But while the looming nuclear catastrophe had relatively few physical effects and has left few physical traces, the consequences of global warming are lasting and implacable.

We are facing a challenge that forces us to ask three questions: What do we know about it? What should we do about it? Why are we doing so little? These are the questions for science, ethics, and a theory of technology.

Science has attended its task comprehensively and rigorously. Comprehensively—that means over a long period of investigation and through the cooperation of many disciplines. In the early nineteenth century, Joseph Fourier discovered the significance of the atmosphere for the temperature of the planet. In the late nineteenth century, Svante Arrhenius investigated the rise of global warming due to increasing greenhouse gases. Once the conjecture of global warming had become plausible, historians began gathering temperature records as far back and as wide across the earth as possible. Biologists turned to dendrochronology, chemists to air bubbles in ancient ice, ecologists to satellite data, etc.

The scientific approach has been rigorous because it did not remain at the level of conjectures, intuitions, and anecdotes, but collected and collated quantitative data. Scientists also attended rigorously to the limits and uncertainties of the data by employing probability and game theories that cast light on the likelihood of developments, the effects of interventions, and the interplay of probabilities and utilities.

Science has proceeded not only comprehensively and rigorously, but forensically as well, i.e., through a public process of conjectures and refutations that has winnowed the wheat from the chaff. Humans are fallible, and so are scientists. When they deal with as complex an issue as global climate, missteps are inevitable. Human fallibility, moreover, is emotional as well as cognitive, and hence ambition, jealousy, and sloppiness at times intrude on research and fuel the glee of climate skeptics. Yet over the long run, the scientific facts have been emerging, and they are distressing. Global warming is real and progressive and the consequences of further warming will be catastrophic for many plants, animals, and above all humans.

So what should we do about global warming? Given what *is* the case, what follows for what we *ought* to do? Science attends to the *is*, ethics to the *ought*. Ethics, no less than science, is a scholarly discipline that has its own theories. Ethics, of course, can also be a practice, the habitual way we do things. It's a well-established principle of western culture that theory aids practice, and when it comes to global warming, our habitual procedures have been so thoughtless that theoretical illumination seems eminently reasonable and desirable. Are the ethical theories equal to the challenge of global warming? What, to start with, are the major ethical theories? By common scholarly consensus, there are three: (1) The ethics of rights and liberties (called deontological ethics by professionals); (2) the ethics of pleasure and prosperity (called utilitarianism); and (3) the ethics of moral excellence (virtue ethics, so called).

Immanuel Kant (1724–1804) is the great historical proponent of deontological ethics. John Rawls (1921–2002) is its most prominent contemporary representative. The epochal achievement of deontological ethics has been the articulation of equal human rights. The immediate consequence of that accomplishment for global warming is the moral requirement that the burdens of global warming be distributed equally or at least fairly. It's less obvious what the ethics of rights and liberties has to say about how we should treat not just humans but also the natural environment. As long as equality and sustainability are embraced, does it matter how we deal with nature? In 1972 Christopher Stone extended the ethics of rights and liberties to nature by giving an affirmative answer to the question that titled his essay: Do trees have standing? Some of the major figures in environmental ethics, Holmes Rolston and Baird Callicot among them, have supported and elaborated this position under the heading of the intrinsic value of nature that humans are morally required to respect. There is a yet further extension of deontological ethics to nature. Biocentrists claim that the equality of rights extends to all living beings; human rights do not trump the rights of nature.

The second ethical theory, utilitarianism, holds that we are morally required to maximize pleasure. The difficult question is: Whose pleasures? John Stuart Mill (1806–1873) is the classic expositor of utilitarianism, and he already included all sentient beings in the population whose pleasures are to be maximized. Peter Singer is the contemporary philosopher who is best known for advocating the promotion of pleasure for all beings capable of experiencing pleasure. An important feature of Singer's view is his requirement that resources be used where they generate most pleasure. It follows, problematically for most people, that we should favor healthy animals over incurable humans and, commendably, that we should use discretionary resources for the poor around the globe rather than for the overfed and overgratified in the advanced industrial countries.

Aristotle (384–321 BCE) is the founder of the third ethical theory, the ethics of moral excellence. It's been revived in our time by Philippa Foot (1920–2010) and Alastaire McIntyre. Here the moral requirement is to pursue moral excellence by acquiring and practicing virtues such as wisdom, courage, friendship, and justice. Virtue ethics has no obvious consequence for how we should respond to global warming except for the requirement that virtue ethics shares with deontological ethics, viz., that we should do so fairly. A second consequence perhaps is the inference that the neglect and abuse of the natural environment is unbecoming to the morally excellent person. It's possible of course to extend the list of virtues to include environmental virtues, and Ronald Sandler has done just that.

There has been much disputing among philosophers which of the three theories is best, and there has been disagreement and rivalry even within the subfield of environmental ethics. As Andrew Light has urged, such squabbles keep philosophers from addressing the environmental crisis publicly and effectively. In fact, a shared and urgent philosophical message is not hard to summarize: Deal with global warming in a way that meets the requirements of global justice and environmental respect.

Science warns and threatens us; ethics admonishes and scolds us, as they should. But it's all been relatively ineffective. Why are we in the United States doing so little to slow and eventually stop our contribution to global warming? The McKinsey report of 2007, "Reducing US Greenhouse Gas Emissions: How Much at What Cost?" answered its question this way: Much at little cost. On the assumption that there would be "no material changes in consumer utility or lifestyle preferences," the United States "*could reduce greenhouse gas emissions in 2030 by 3.0 to 4.5 gigatons of* CO_2*e* [carbon dioxide equivalents] *using tested approaches and high-potential emerging technologies.*" The US emissions in 2008 were 6.957 CO_2e.

The most ready answer to why so little is being done is to assume that the citizens of this country are ignorant of the scientific facts and indifferent to the claims of ethics. But why are they ignorant? And why are they indifferent? There is a morally honorable answer to the first question if there are good reasons why people are ignorant, and a like answer is available to the second question if there are good reasons why people are indifferent.

People have reason to be ignorant if literacy is unobtainable or information unavailable. But neither is the case for most people in this country. People have reason to be indifferent to other people's misery if they are struggling with mortal misery themselves. But that's not the case either for most Americans. We're inclined to think of people who are inexcusably ignorant and indifferent as indolent. But for the most part Americans are decent people. We all know this anecdotally and from first-hand observation. People show up for work, do a good job, are helpful when asked for directions, and will give you change if you need coins for a parking meter.

General decency is borne out by polls, most tellingly for our purposes when they reveal a combination of decency and ignorance. In 2001, Martin Gilens found that people wrongly assumed that 25 percent of the Federal Budget was spent on foreign aid. That was ignorant. The right amount, people thought, would be ten percent which would in fact have been fifteen times what we actually spent at the time. That was decent. In 2010, Michael Norton and Dan Ariely found that people assumed the top quintile of the population owned 60 percent of total wealth when their share was in fact 85 percent. That was ignorant. A fair share for the top quintile, people thought, would be 35 percent. That was decent.

The resolution of the indolence-decency paradox comes into view when we take an incisive look at the cultural structure of contemporary reality. It is ruled by a system, called late capitalism by Marxists, the consumer society by pundits, and technology by Heidegger. The Heideggerian term draws most helpfully on our commonsense intuitions, and it's the one I'm adopting here.

We spend our daily lives within the technological system as we must. What we have yet to learn is to see the system from without and at some distance. Living within technology, we are subject to its constraints. They can be divided into social and material constraints, and both range from hard and universal to soft and partial. We are much more aware of the social constraints—the laws, the regulations, the codes of courtesy, the philosophies of life, and religious doctrines. There are vigorous and expansive discussions of the rise and fall of these constraints and of their conflicts and compatibilities.

It's the material constraints that we are by and large unaware of although they are morally as potent as the social ones. Being for the most part unaware of them, we largely fail to take responsibility for them. Like the social constraints, the material ones extend from hard to soft. Getting our heat and water from utilities is a hard and all but universal constraint. Owning a car is a little softer and less universal. Getting our food from supermarkets is less coercive yet. Having access to the Internet is a requirement that is hardening as I write.

These observations are, once you think about them, reminders of the obvious. What needs special attention and analysis is the peculiar quality of the soft material constraints, what we may call the inducements of the system of technology. Consider the inducements to eat. They are ubiquitous, attractive, and easily available. The result is that two thirds of the population is helplessly exposed to too much food.

The paradigm of instantaneous, ubiquitous, and attractive availability is a crucial feature of the rule of technology. This sort of availability has crowded out all prior and competing engagements with reality with the crucial exception of work. Everyone understands that technology is not magic and that the commodities of consumption rest on a machinery that needs to maintained, repaired, and improved. All of us realize that we have to pay our dues to the machinery of technology. There is then a common and implicit grasp of the system of technology. But there is no lucid and responsible comprehension of its structure and moral potency, and it is that lack of insight that leaves the findings of science and the exhortations of ethics without much consequence.

To begin with science, immersion in scientific reports and appropriation of scientific information into politically influential knowledge seems like a pointless and tedious enterprise in a world where information has become an article of easy and attractive consumption. As for ethics, it is not surprising that the ethics of rights and liberties has been the most effective since it deals with the social constraints in the system of technology. But that strength is also its weakness—its obliviousness to the moral implications of the material culture.

Regarding utilitarianism, the morally admirable version of Peter Singer is overwhelmed by its mainstream twin that takes the self-interested maximizing of pleasure for its guiding norm and has mostly been unwilling to protest the equation of pleasure with consumption. Virtue ethics, finally, is unlikely to have a purchase on people who are swamped by the inducements within the system of technology. Wisdom must seem like an incomprehensible abstraction. Courage is irrelevant when none of the paradigmatic pleasures requires discipline and daring. And friendship is being subverted and commodified by social networks.

I have overdrawn the picture of life within technology. It is important for a theory of technology to expose its injurious features that confront us in daily episodes as well as in the findings of the social sciences. It is just as important to show that the rule of technology is not total. It has left openings of engagement with reality—the culture of the table, the excursions into nature, the practices of the arts, and for some of us the consolations of poetry or of religion.

Just as the paradigmatic pattern of technology has escaped sustained public scrutiny, so the focal things and practices that can sustain us have lacked public recognition and support. But they provide the footholds that are needed to gain some distance on the system of technology, to see its blessings and burdens, and to take responsibility for that system, most to point here, for the burdens of global warming and global injustice.

Scientists have to continue to warn and threaten us. The advocates of global justice and environmental stewardship must persevere in scolding and exhorting us. But in addition thoughtful people need to understand and expose the powerful constraints of technology as well as the constructive alternatives of engaging things and practices. This additional and complementary view of the world invites us to go beyond warning and scolding. To urge people to shift from consumption to engagement and from indolence to responsibility is to be a bearer of good news.

12.

TEN ETHICAL QUESTIONS THAT SHOULD BE ASKED OF THOSE WHO OPPOSE CLIMATE CHANGE POLICES ON SCIENTIFIC GROUNDS

Donald A. Brown

I. INTRODUCTION

Those who oppose proposed legislation or policies to reduce the threat of climate change often base their opposition on scientific uncertainty or claims that there is no scientific basis for conclusions that human activities are dangerously causing climate change. These arguments range from assertions that what is usually called the "mainstream" scientific climate change view is a complete hoax to the milder assertions that the harsh climate change impacts on human health and the environment predicted by the Intergovernmental Panel on Climate Change (IPCC) and other climate change researchers are unproven.

In responses to the lack-of-scientific-proof arguments, climate change policy advocates usually stress the harsh environmental impacts to people and ecosystems that climate change will cause if action is not taken or argue that climate change science is settled. In other words, advocates of climate change action respond to claims of opponents to climate change programs by denying the factual claims of the opponents and by making other factual claims.

After a short review of why the consensus view among climate change scientists is entitled to respect, the chapter will argue that decision making in the face of uncertainty about climate change impacts must be understood as raising ethical questions, not only "value-neutral" scientific "factual" questions, even though scientific understanding of potential climate change impacts is relevant to the ethical issues.

In looking at these issues, the chapter will review some of the most recent arguments made against climate change policies on scientific grounds in the United States, including issues raised by hacked e-mails of mainstream climate change scientists. The chapter will then explain why decision making in the face of uncertainty must be understood to raise ethical questions.

This chapter will identify ten ethical questions that should be asked of those who oppose climate change policies on scientific grounds. These questions will be shown to be relevant to specific arguments recently made by opponents of climate change

The chapter has the following sections:

II. The Consensus View
III. Arguments Made against Climate Change Policies on the Grounds of Scientific Uncertainty
IV. The Email-Gate Scandal
V. Why Decisions in the Face of Uncertainty Must Be Understood as Raising Ethical Questions
VI. Ethical Questions That Should Be Asked

II. THE CONSENSUS VIEW

Before examining arguments made against climate change policies on the basis of scientific uncertainty, the chapter will first review what is frequently called the scientific "consensus" position. The consensus view is usually understood to be that articulated by the Intergovernmental Panel on Climate Change.[1]

The IPCC was established by the World Meteorological Organization (WMO) and the United Nations Environment Programme (UNEP) in 1988 to assess for governments the scientific, technical, and socioeconomic information relevant for the understanding of climate change and its potential impacts and consequent options for adaptation and mitigation.[2] The IPCC does not do original research but synthesizes and summarizes the peer-reviewed climate change science for governments.[3]

Any government that is a member of the World Health Organization (WHO) or UNEP may be a member of the IPCC. The coordinating work of the IPCC is conducted by the IPCC general assembly, where every member country has one vote. Therefore governments that have often opposed international action on climate change, such as Saudi Arabia, have the same power as governments that have traditionally supported international action, including many of the small, developing island states like Tuvalu. The assessment reports of the IPCC have been unanimously approved by the member countries, including, for instance, by the United States and Saudi Arabia, two countries that have, for most of the history of international negotiations, opposed establishing the international climate change regimes that most observers believe are necessary to prevent dangerous climate change.

The first IPCC assessment report was published in 1990; the second in 1996; the third in 2001; and the fourth in 2007. Each IPCC report drew conclusions linking human activities to observable warming with increasing levels of certainty.[4] The IPCC shared the 2007 Nobel Peace Prize with former vice president of the United States Al Gore.

The Fourth Assessment Report (AR4) was completed in early 2007. Like previous reports, this assessment consists of four reports, three of them from each of its working groups. Working Group I deals with the physical science basis for climate change; Working Group II assesses climate change impacts; and Working Group III assesses options for mitigating climate change through limiting or preventing greenhouse gas emissions and enhancing activities that remove them from the atmosphere.[5] In addition to the reports of the working groups, AR4 also contained a Synthesis Report.[6]

The Working Group I Summary for Policymakers (SPM) drew the conclusion that human actions were causing dangerous climate change with higher levels of certainty than in previous reports. Its key conclusions were:

- Warming of the climate system is unequivocal.
- Most of the observed increase in globally averaged temperatures since the mid-twentieth century is very likely due to the observed increase in anthropogenic (human) greenhouse gas concentrations.
- Anthropogenic warming and sea level rise would continue for centuries due to the time scales associated with climate processes and feedbacks, even if greenhouse gas concentrations were to be stabilized, although the likely amount of temperature and sea level rise would vary greatly depending on the fossil intensity of human activity during the next century.
- The probability that this is caused by natural climatic processes alone is less than 5 percent.
- World temperatures could rise by between 1.1 and 6.4 °C (2.0 and 11.5 °F) during the twenty-first century and, as a result:
    - Sea levels will probably rise by 18 to 59 cm (7.08 to 23.22 in.)
    - There is a confidence level > 90 percent that there will be more frequent warm spells, heat waves, and heavy rainfall.
    - There is a confidence level > 66 percent that there will be an increase in droughts, tropical cyclones, and extreme high tides.
- Both past and future anthropogenic carbon dioxide emissions will continue to contribute to warming and sea level rise for more than a millennium.
- Global atmospheric concentrations of carbon dioxide, methane, and nitrous oxide have increased markedly as a

> result of human activities since 1750 and now far exceed preindustrial values over the past 650,000 years[7]

As we shall see, the vast majority of climate scientists and scientific organizations agree with the consensus position articulated by the IPCC. Yet, some skeptical scientists, many of whom are associated with conservative think tanks, have frequently made criticisms of the IPCC's conclusions. These criticisms are of many types and range from claims that the IPCC is overestimating adverse climate change impacts to assertions that there is no evidence that observed warming is attributable to human actions.[8] Responses to these arguments from the mainstream scientific view are widely available.[9]

Recent reports have concluded that the vast majority of scientists actually doing research in the field support the consensus scientific view. For example, a 2009 study—published in the *Proceedings of the National Academy of Sciences of the United States*—polled 1,372 climate researchers and resulted in the following two conclusions:

> 1. 97–98 percent of the climate researchers most actively publishing in the field support the tenets of ACC (Anthropogenic Climate Change) outlined by the Intergovernmental Panel on Climate Change, and
> 2. the relative climate expertise and scientific prominence of the researchers unconvinced of ACC are substantially below that of the convinced researchers.[10]

Another poll performed in 2009 from 3,146 of the 10,257 polled earth scientists concluded that 76 out of 79 climatologists who "listed climate science as their area of expertise and who also have published more than 50 percent of their recent peer-reviewed papers on the subject of climate change" believe that mean global temperatures have risen compared to pre-1800s levels, and 75 out of 77 believe that human activity is a significant factor in changing mean global temperatures.[11]

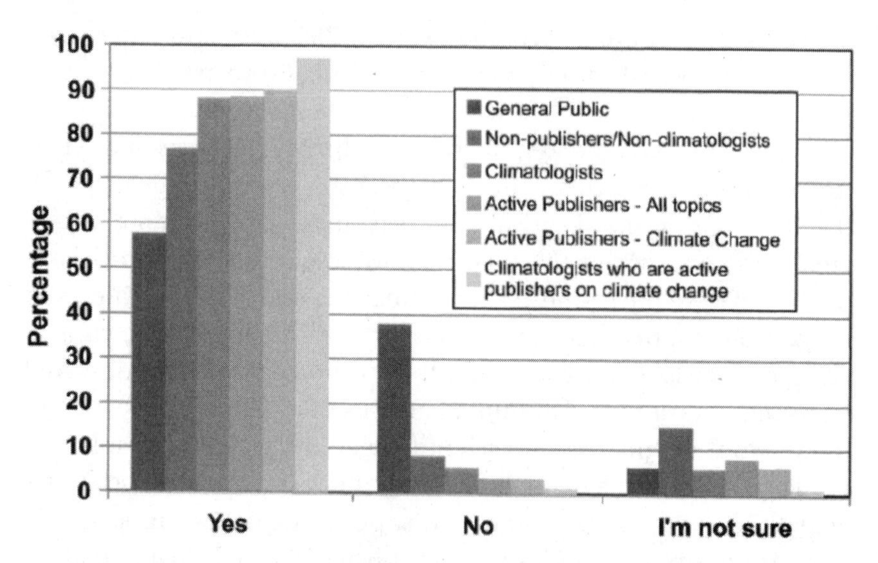

Responses to poll question: ***Do you think human activity is a significant contributing factor in changing the climate?***[12] (Copyright 2009 American Geophysical Union. Reproduced by permission of American Geophysical Union.)

In response to arguments from climate change skeptics, many scientific organizations with expertise relevant to climate change have endorsed the consensus position that "most of the global warming in recent decades can be attributed to human activities," including the following:

- American Association for the Advancement of Science
- American Astronomical Society
- American Chemical Society
- American Geophysical Union
- American Institute of Physics
- American Meteorological Society
- American Physical Society
- Australian Coral Reef Society
- Australian Meteorological and Oceanographic Society
- Australian Bureau of Meteorology and the Commonwealth Scientific and Industrial Research Organization

- British Antarctic Survey
- Canadian Foundation for Climate and Atmospheric Sciences
- Canadian Meteorological and Oceanographic Society
- Environmental Protection Agency
- European Federation of Geologists
- European Geosciences Union
- European Physical Society
- Federation of American Scientists
- Federation of Australian Scientific and Technological Societies
- Geological Society of America
- Geological Society of Australia
- International Union for Quaternary Research
- International Union of Geodesy and Geophysics
- National Center for Atmospheric Research
- National Oceanic and Atmospheric Administration
- Royal Meteorological Society
- Royal Society of the United Kingdom[13]

The academies of science from nineteen different countries all endorse the consensus. Eleven countries have signed a joint statement endorsing the consensus position. They are:

- Academia Brasiliera de Ciencias (Brazil)
- Royal Society of Canada
- Chinese Academy of Sciences
- Academie des Sciences (France)
- Deutsche Akademie der Naturforscher Leopoldina (Germany)
- Indian National Science Academy
- Accademia dei Lincei (Italy)
- Science Council of Japan
- Russian Academy of Sciences
- Royal Society (United Kingdom)
- National Academy of Sciences (United States)[14]

From this it can be seen that the debate about the consensus view articulated by the IPCC is strongly supported by the vast majority of climate change scientists. In fact, some critics have contended that the IPCC reports tend to underestimate dangers and understate climate change risks because the process that leads to the IPCC conclusions gives representatives from countries that have consistently opposed the adoption of international climate regimes power to pressure the IPCC scientists to report only the lowest common denominator. In fact observations of actual greenhouse gas (ghg) atmospheric concentrations, temperatures, and sea level rise are near or exceeding IPCC worst-case predictions. One recent comparison of ghg concentrations, temperatures, and sea-level-rise observations versus predictions concluded:

> Overall, these observational data underscore the concerns about global climate change. Previous projections, as summarized by IPCC, have not exaggerated but may in some respects even have underestimated the change, in particular.[15]

III. ARGUMENTS MADE AGAINST CLIMATE CHANGE POLICIES ON THE GROUNDS OF SCIENTIFIC UNCERTAINTY

The threat of climate change was recognized by US politicians as early as President Lyndon Johnson's February 1965 special message to Congress that stated:

> This generation has altered the composition of the atmosphere on a global scale through radioactive materials and a steady increase in carbon dioxide from the burning of fossil fuels.[16]

International interest in reducing greenhouse gas emissions grew dramatically in the late 1970s as computer modelers began to use new computing tools to construct climate models that were capable of predicting temperature changes caused by human-induced climate change. In 1977, Robert M. White, the head of the National Oceanic and Atmospheric Administration wrote a report for the National Research Council that concluded that CO_2 released during the burning of fossil fuel can have consequences for climate that pose a considerable threat to future society.[17]

A report prepared by the Carter administration in 1981 declared that "the responsibility of the carbon-dioxide problem is ours—we should accept it and act in a way that recognizes our role as trustees for future generations."[18] This report also estimated that the amount of warming that would be experienced from a doubling of preindustrial levels of CO_2 would be 3 °C, very close to the amount that the IPCC would predict almost thirty years later.[19] However, global warming was not a priority of the successor Reagan administration although international interest in climate change grew rapidly in the 1980s.[20]

During the 1980s according to Naomi Oreskes and Erik Conway, several economists writing reports for the Reagan administration concluded that not much could be done to avoid the threat of climate change and advised going slow for economic reasons and because of uncertainties about climate change impacts, and in so doing ignoring what other scientists were saying.[21]

Also during the 1980s, a few scientists that produced bogus information on behalf of the tobacco industry about the risks of smoking began arguing in reports issued by the George C. Marshall Institute that the sun rather than human burning of fossil fuels was responsible for recent warming.[22] These claims would be reviewed by the IPCC and rejected on the basis that if the sun were the source of the warming, the upper atmosphere would warm at the same time the lower atmosphere heated. Yet the upper atmosphere had been cooling as the planet warmed.[23] Attributing warming to the sun is one of many arguments that have been made by skeptics in the last twenty years that have been examined and refuted by scientists that support the consensus position on the issue of whether humans are responsible for observed warming (see discussion below on fingerprinting).

Also beginning in the 1980s, a number of corporate-funded campaigns were created to convince Americans that concerns about potential adverse effects of human-induced climate change were not based upon sound science.[24] These campaigns sometimes were run by public relations firms who had been hired by corporations or industry associations that had financial interests in cheap fuel. For instance, Western Fuels, the National Coal Association, and the Edison Electric Institute created the Information Council on the Environment (ICE) to "reposition global warming as a theory (not fact)" and "supply alternative facts to support the suggestion that global warming will be good."[25] ICE worked with the public relations firm Bracy Williams and Company to convince the public that human-caused climate change may not exist.[26]

Among many organizations financed by corporations to fight climate change policies in the 1990s was the Global Climate Coalition, an organization whose members primarily were heavy producers of greenhouse gases, mostly from the United States.[27] The Global Climate Coalition's budget in 1997 alone was $1.68 million, and up until 2002 this organization worked to convince public officials and civil society that climate change was not proven and therefore government climate change policies were not necessary.[28] The major funders of the Global Climate Coalition were ExxonMobil, Royal Dutch Shell, British Petroleum (now BP), Texaco, General Motors, Ford, DaimlerChrysler, the Aluminum Association, the National Association of Manufactures, the American Petroleum Institute, and others.[29] The Global Climate Coalition was formed in 1989 but became defunct in 2002 after several of its members withdrew because of a flap triggered by an April 2009 *New York Times* report that stemmed from a document revealed in court as part of a lawsuit. The document demonstrated that "even as the coalition worked to sway opinion, its own scientific and technical experts were advising that the science backing the role of greenhouse gases in global warming could not be refuted.[30] In other words, the Global Climate Coalition sought to undermine the consensus position on climate change even though they had been advised that the consensus position was entitled to respect by their own experts.

To convince Americans that the IPCC conclusions on climate change were not scientifically supportable, Republican pollster Frank Luntz advised Republican politicians in a memo to continue to make scientific uncertainty a primary issue in the debate, and refer to other scientists in the field.[31] In this memo, Luntz also advised Republicans that "you need to be more active in recruiting experts who are sympathetic to your view, and much more active in making them part of your message."[32]

During the 1980s up to the present, many conservative think tanks were created for the express purpose of defending free-market capitalism.[33] These think tanks did not produce "objective" science but were focused upon producing skepticism about environmental problems.[34] They acted as strategists and lobbyists in the war of ideas to protect free enterprise against government regulation. The number of think tanks in the United States between 1970 and 1996 grew from fewer than sixty to well over three hundred.[35] The conservative think tanks were funded primarily by large businesses and free-market foundations. Although the think tanks that first appeared in the United States in the 1950s were initially nonideological, the last thirty years has seen enormous growth in think tanks with a clear ideological mission.[36]

It is clear from the mission statements of many of the conservative think tanks that they are not likely to produce research that leads to the conclusion that human activities threaten human health and the environment and therefore should be regulated. It is also clear from examining the sources of funding of these think tanks that their very existence depends upon funding from organizations and corporations that are hostile to environmental regulation. Yet the research and analysis produced by these think tanks is often treated by the media as if it were value-free objective analysis of policy options.

Although they are producers of research and policy analyses, think tanks don't usually subject their work to peer review or diversity-of-thought checks that are the norm in research produced by academics. Yet these think tanks implicitly or explicitly claim that their policy analyses are scientifically sound and unbiased. However, it is clear that the ideological think tanks generate analyses that promote the interests of their financial sponsors; that is, right-wing philanthropic organizations or corporations.

When these think tanks release reports on policy that are reported on by the press, the ideological focus of these institutions is rarely identified.[37] For instance in 1997 only 14 percent of the 132 stories sampled in which the American Enterprise Institute (AEI) was mentioned identified it as conservative.[38] Cato was similarly not labeled in 68 percent of the 130 stories sampled. It was identified as "libertarian" 13 percent of the time, "conservative" 6 percent of the time, and twice was referred to as both "libertarian" and "conservative." One reference called the institution "free-market oriented."[39]

A few fossil fuel companies have used right-wing think tanks as tools to prevent regulation of their products to reduce greenhouse gas emissions. According to *Mother Jones* magazine, ExxonMobil has funded forty think tanks and media outlets to preach skepticism about whether climate change will create serious problems.[40] These think tanks have made arguments such as:

- The science of climate change has been debunked;
- Global warming can actually save lives;
- Not only is the scientific basis of global warming increasingly uncertain, but Kyoto will also ultimately prove to be an economic disaster for Europe—and the developing world;
- The science behind global warming is inconclusive, and to teach otherwise is fearmongering;
- Recent evidence shows that global warming may not be happening;
- The scientific hypothesis for global warming is scientifically weak;
- No one seriously claims to know whether past warming was caused by human activities; whether further warming will occur and, if it does, whether it will result from human activities; and whether such warming in some general sense would be a bad thing;
- The costs to the United States economy of dealing with climate change are too great to justify action.[41]

The think tanks funded by ExxonMobil include some of the largest think tanks as well many smaller think tanks that appear to have been created for the express purpose of influencing climate change policy.

The magnitude of the incongruence between the scientific consensus position on climate change held by most climate scientists and the arguments made by the think tanks is large. Yet the think tanks have been successful in convincing many in the United States that climate change may not be a threat. This is so despite the growing realization in the scientific community of the enormity of the threat to human health and the environment from climate change due to rising seas, increases in storm damage, vector-borne disease, droughts and floods, loss of biodiversity, heat deaths, and increased water scarcities. In addition, those who will most likely be harmed most by climate change are some of the poorest people around the world, particularly the millions of people vulnerable to rising seas, droughts, and floods.

In November 2006 it was discovered that Kenneth Green, a resident scholar from the American Enterprise Institute, was offering cash in the amount of $10,000 plus expenses to any scientist who would write a critique of the then upcoming Fourth IPCC Assessment.[42] Long before the IPCC report was issued, apparently AEI had decided that it would not like the IPCC results and was looking for scientists who would help criticize the findings whatever they would be. AEI had received significant corporate funding.[43]

In addition to corporate funding, the conservative think tanks receive most of their funding from mostly conservative foundations that have had as part of their mission protecting free-market capitalism. They are:

> The Lynde and Harry Bradley Foundation, the Carthage Foundation, the Earhart Foundation, the Charles G. Koch, David H. Koch and Claude R. Lambe charitable foundations, the Phillip M. McKenna Foundation, the JM Foundation, the John M. Olin Foundation, the Henry Salvatori Foundation, the Sarah Scaife Foundation, and the Smith Richardson Foundation.[44]

And so, many, although not all, arguments made against governmental climate change policies on the basis of scientific uncertainty have been orchestrated by climate skeptics associated with think tanks whose very mission is hostile to government regulation. For this reason it can be concluded that not all arguments skeptical of the consensus view about the dangers of climate change are entitled to respect.

IV. THE EMAIL-GATE SCANDAL

In November 2009 a new controversy about the science of climate change broke out with the publishing of thousands of hacked e-mails and other documents from the University of East Anglia's (UEA) Climatic Research Unit (CRU). This scandal was quickly labeled the "EmailGate" scandal. This scandal received huge public attention around the world right before the international community was scheduled to meet in Copenhagen for international climate change negotiations.

The EmailGate controversy quickly received much global attention even though the documents did not seriously threaten the mainstream scientific consensus view about climate change.[45] The hacked e-mails were widely circulated on the Internet and in the media by climate skeptics often associated with conservative think tanks who charged that the e-mails proved that the consensus view on climate change was manufactured or even perhaps fraudulent. However, since then a strong scientific consensus has arisen that the e-mails don't lead to the conclusions that have been claimed by the skeptics; that is, they don't provide proof that human-caused climate change is a lie or a swindle.[46]

For a while, the media treated the EmailGate controversy as if it disproved the consensus view on climate change. According to Fred Peirce from the *Guardian*,

> Almost all the media and political discussion about the hacked climate emails has been based on soundbites publicized by professional skeptics and their blogs. In many cases, these have been taken out of context and twisted to mean something they were never intended to.[47]

Recently, *Newsweek* summarized the EmailGate controversy this way:

> A lie can get halfway around the world while the truth is still putting its boots on, as Mark Twain said (or "before the truth gets a chance to put its pants on," in Winston Churchill's version), and nowhere has that been more true than in "climategate." In that highly orchestrated, manufactured scandal, e-mails hacked from computers at the University of East Anglia's climate-research group were spread around the Web by activists who deny that human activity is altering the world's climate in a dangerous way, and spun so as to suggest that the scientists had been lying, cheating, and generally cooking the books.[48]

Several investigations have now exonerated Phil Jones from the University of East Anglia and Michael Mann from Penn State University, the main protagonists in the skeptics' accusations that the emailgate controversy proved that the mainstream climate change scientists had hidden scientific information inimical to the positions they were taking.[49]

Yet newspapers like the *Wall Street Journal* and US politicians such as James Imhoff, who had previously claimed that global warming was "the greatest hoax ever perpetrated on the American people," have asserted that EmailGate evidence supports the conclusion that the IPCC conclusions about human-induced warming are not sound.

At the center of the skeptics' assertions about the meaning of the hacked e-mails were claims made about Michael Mann's hockey stick, a graph prominently displayed by the IPCC, which was created both from reconstructions of historical climate change and the more recent actual instrumental record. The skeptics used the EmailGate controversy to discredit the IPCC conclusions about human-induced climate change and, in so doing, often made central an e-mail from Phil Jones that was claimed to acknowledge that the hockey stick was constructed to "hide the decline" in temperatures that is indicated by tree-ring data. Yet recent exonerations of Jones and Mann have concluded that there is no evidence that either scientist tried to deceive the public about the implications of the tree-ring data and that "hide the decline" was a term commonly used in the science to display data where one source of the data was known to be unreliable. Of greater significance in regard to undermining the consensus position, skeptics who publicized the hacked e-mails and the "hide the decline" e-mail in particular to undermine IPCC conclusions that human actions were causing climate change ignored the fact that the hockey stick was never used as the basis for attributing global warming to human actions. One mainstream climate change scientist recently summed up the situation as follows:

> Unfortunately, Climate Denier Gate is being used by some on blogs and the usual suspects in media to lionize the perpetrators as some kind of heroic stealth investigative reporters who have just in time saved the world from the "big mistake" of fashioning climate policy in Copenhagen. The amazing scientific thing that nobody seems to be covering is that the "hockey stick" was never used as proof of anthropogenic global warming by IPCC—it was the "fingerprinting" studies of many scientists dating back to 1995—three years before the first hockey stick was even published. A fingerprint is an attempt to combine models of climate change with observed data. . . . That fingerprint history the denier set will almost certainly not mention, just claim that the hockey stick guys are "exposed" and therefore AGW (anthropgenic global warming) is a fraud. The fraud however is on the deniers, I'm afraid, since the hockey stick has (a) never been disproved, and (b) nor was it ever the basis for AGW; likelihood assessment. Rather, the fingerprint analyses by many groups over the years were the scientific evidence used for AGW. Would somebody in the mainstream media please cover this!!![50]

The mainstream scientific consensus on climate attributes most of the undeniable warming that the world is experiencing to human causation. This attribution is based largely on how the planet warms up differentially depending upon whether it is caused by increases in greenhouse gases or forcing that comes from natural causes. This is called "fingerprinting." If the upper atmosphere warms as the lower atmosphere cools, if the nights warm up faster than days, if oceans warm at depth, if the boundary between the lower and upper atmosphere known as the "tropopause" moves higher, if the amount of heat escaping the atmosphere decreases, since all of these phenomenon are expected by increases in greenhouse gases rather than other processes that drive climate change, then there is a strong scientific basis for concluding that human increases in greenhouse gas emissions are the cause of warming.[51] Since all of these phenomenon are being observed, this is strong basis for human attribution as the cause of climate change. The EmailGate scandal had nothing to do with these issues on which much of the scientific consensus rests. Yet many of the climate change skeptics and the skeptical climate change think tanks claimed that the EmailGate controversy demonstrated that the consensus position was a hoax.

Given this, many scientific organizations with expertise on climate change science have now issued statements expressly affirming that EmailGate has not dented the scientific basis on which the consensus position rests.

- The IPCC has issued a statement asserting that despite EmailGate it strongly supports its conclusions in the fourth assessment released in 2007.[52]
- The American Meteorological Society stated that EmailGate did not affect the society's position in support of the consensus view on climate change.[53]
- The American Geophysical Union (AGU) stated that they found "it offensive that these emails were obtained by illegal cyber attacks and they are being exploited to distort the scientific debate about the urgent issue of climate change." They reaffirmed their 2007 position statement on climate change "based on the large body of scientific evidence that Earth's climate is warming and that human activity is a contributing factor."[54]
- The American Association for the Advancement of Science (AAAS) reaffirmed its position on global warming and "expressed grave concerns that the illegal release of private emails stolen from the University of East Anglia should not cause policy-makers and the public to become confused about the scientific basis of global climate change."[55]
- The UK House of Commons Science and Technology Select Committee inquiry reported on March 31, 2010, that it had found that "the scientific reputation of Professor Jones and CRU remains intact and the allegations against him were baseless."[56]

- An independent scientific review committee headed by Sir Muir Russel exonerated the scientists caught up in Climate-Gate in a lengthy report.[57]
- Penn State University prepared two reports completely exonerating Michael Mann, one of the scientists at the center of the EmailGate controversies.[58]

And so, we have seen for almost three decades that climate change deniers have made scientific arguments in opposition to proposals of national governments to limit greenhouse gas emissions and international climate change regimes. Initially some of the scientific skeptics on climate change had previously worked on denying the connection between tobacco and cancer. We have also seen that in the last thirty years many right-wing think tanks have been funded by corporations with financial interests in minimizing the scope of climate change regimes or conservative foundations with a free-market agenda. These think tanks have often worked with climate change deniers to convince civil society and politicians that climate change offers no threat.

Although the vast majority of scientists that actually publish in the field of climate change science support the consensus position, not all do. Because not all climate change scientific issues have been resolved, even if one assumes that the consensus position articulated by the IPCC is entitled to respect, and because the advance of science depends upon probing scientific claims skeptically, skepticism about climate change claims is warranted. Doubt is necessary for science to progress. Yet doubters are expected to expose their doubts to peer review, a process whereby claims are subject to critical scrutiny. Many of the assertions of climate skeptics were never subjected to peer review scrutiny. According to a paper published in *Science* by Naomi Oreskes, out of 928 articles published in prominent scientific journals between 1993 and 2003, none disagreed with the conclusion that human activities were causing climate change.[59] As we shall see, as a matter of ethics, skeptical claims should also be questioned about whether they are asserting that they have proven that human releases of greenhouse gases are safe.

V. WHY DECISIONS IN THE FACE OF UNCERTAINTY MUST BE UNDERSTOOD AS RAISING ETHICAL QUESTIONS

As we have seen, many scientific arguments against climate change policies implicitly argue that climate change policies should be opposed because they are not in the US national interest. In other words, if the United States adopts climate change policies that turn out to be costly but unnecessary to prevent climate change, the US economy is harmed.

The responses of advocates of US climate change policies to these arguments are almost always to take issue with the factual economic and scientific conclusions of these arguments by making counter economic and scientific claims. For instance, in responses to the lack-of-scientific-proof arguments, climate change advocates usually stress the harsh environmental impacts to people and ecosystems that climate change will cause if action is not taken or argue that climate change science is settled. In other words, advocates of climate change action respond to claims of opponents to climate change programs by denying the factual scientific claims of the opponents.

By simply opposing the factual claims of the opponents of climate change, the advocates of climate change policies are implicitly agreeing with the assumptions of their opponents that greenhouse reduction policies should not be adopted if they are not in the national self-interest.

Yet climate change is a problem that clearly creates civilization-challenging ethical issues. By ethics is meant the domain of inquiry that examines claims that, given certain facts, actions are right or wrong, obligatory or nonobligatory, or responsibilities attach to human activities.

If nations or individuals have ethical obligations, they are likely to have duties, responsibilities, and obligations that require them to go beyond consideration of self-interest alone in making decisions. And so, if climate change raises ethical considerations, governments may not base policy decisions on self-interest alone.

There are several distinct features of climate change that call for its recognition as creating civilization-challenging ethical questions.

First, climate change creates duties because those most responsible for causing this problem are the richer developed countries, yet those who are most vulnerable to the problem's harshest impacts are some of the world's poorest people in developing countries. That is, climate change is an ethical problem because its biggest victims are people who can do little to reduce its threat.

Second, climate change impacts are potentially catastrophic for many of the poorest people around the world. Climate change harms include deaths from disease, droughts, floods, heat, intense storms, damages to homes and villages from rising oceans, adverse impacts on agriculture, diminishing natural resources, the inability to rely upon traditional sources of food, and the destruction of water supplies. In fact, climate change threatens the very existence of some small island nations. Clearly these impacts are potentially catastrophic.

The third reason why climate change must be understood to be an ethical problem stems from its global scope. At the local, regional, or national scale, citizens can petition their governments to protect them from serious harms. But at the global level, no government exists whose jurisdiction matches the scale of climate change. And so, although national, regional, and local governments have the ability and responsibility to protect citizens within their borders, they have no responsibility to foreigners in the absence of international law. For this reason, ethical appeals are necessary to motivate governments to take steps to prevent great harm to noncitizens.

Climate change scientists around the world have been working to determine the nature of the threat from global warming. From a proposition that a problem like global warming creates a particular threat or risk, one cannot, however, deduce whether that threat is acceptable without first deciding on certain criteria for acceptability. The criteria of acceptability must be understood as an ethical rather than a scientific question. For instance, although science may conclude that a certain increased exposure to solar radiation may increase the risk of skin cancer by one new cancer case in every hundred people, science cannot say whether this additional risk is acceptable because science describes facts and cannot generate prescriptive guidance by itself. The scientific understanding of the nature of the threat, of course, is not irrelevant to the ethical question of whether the risk is ethically acceptable, but science alone cannot tell society what it should do about various threats. In environmental controversies such as global warming where there is legitimate concern, important ethical questions arise when scientific uncertainty prevents unambiguous predictions of human health and environmental consequences. This is so because decision makers cannot duck ethical questions such as how conservative "should" scientific assumptions be in the face of uncertainty or who "should" bear the burden of proof about harm. To ignore these questions is to decide to expose human health and the environment to a legitimate risk; that is, a decision to not act on a serious environmental threat has consequences. Science alone cannot tell us what assumptions or concerns should be considered in making a judgment about potentially dangerous behavior.

* * *

Arguments made to support nonaction on climate change also raise ethical questions because, as we have seen, in at least a few cases, claims have been made by organizations and individual skeptics that have not only been misleading but have been made without regard to the truth. Given what is at stake from climate change, to the extent that these claims amounted to knowing or grossly negligent disinformation, they are grossly immoral.

For this reason, environmental decisions in the face of scientific uncertainty must be understood to raise a mixture of ethical and scientific questions. Yet the scientific skeptics on global warming often speak as if it is irrational to talk about duties to reduce greenhouse gases until science is capable of proving with high levels of certainty what the actual damages will be. The skeptics seem to dismiss the conclusions of the IPCC on the basis that they have not adequately proven that IPCC's identified impacts will happen as described. This condemnation comes despite the fact that the IPCC only claims that their descriptions of global warming impacts are likely or very likely;

that is, they are not proven consequences of the continuing human release of greenhouse gases. The skeptics often attack the scientific proponents of global warming action on scientific grounds, accusing them of doing "bad" science because they rely on unproven assumptions, even though the IPCC's conclusions are based upon its review of peer-reviewed science on global warming. Yet the skeptics not only offer no proof for their alternative predictions, they usually simply attack the assumptions of the mainstream scientists by offering their own unproven and non-peer-reviewed theories about likely timing and magnitude of global warming. By attacking the mainstream scientists' views of likely impacts, the skeptics are implicitly arguing that only proven "facts" should count in the debate. They do this despite the fact that climate models will probably never be able to prove with high degrees of certainty what future temperatures will actually be. This is the case because the climate models will always need to simplify a complex and chaotic climate system, rely on speculation about future population, technology, and use of fossil fuels, and make reasonable guesses about human health and environmental impacts of temperature change through the use of environmental impact science, an inherently uncertain science. Therefore, the skeptics' attack on mainstream climate science on the grounds of its use of unproven assumptions hides a very controversial but unstated ethical position, namely that governments should not act until absolute scientific proof is in. For this reason, the skeptics appear to be opposed to the use of science to describe potentially dangerous behavior. In the case of global warming, the skeptics want only proven science to count in public policy formulation about potentially catastrophic human activities. In taking this position the skeptics are implicitly arguing that the burden of proof should be on those who may be victims of global warming to show that damages to them will actually occur.

And so, if climate change raises civilization-challenging ethical questions that imply duties, responsibilities, and obligations, what questions should the press ask opponents of climate change policies when they make economic and scientific arguments against climate change policies?

VI. ETHICAL QUESTIONS THAT SHOULD BE ASKED

Given that climate change must be understood to raise ethical questions, the press should ask those who oppose climate change policies on scientific grounds the following questions:

1. When you argue that the United States should not adopt climate change policies because adverse climate change impacts have not yet been proven, are you claiming that climate change skeptics have proven that human-induced climate change will not create adverse impacts on human health, the resource base, and the ecological systems of others and, if so, what is that proof?
2. When you argue that the United States should not adopt climate change policies because there is scientific uncertainty about adverse climate change impacts, are you arguing that no action on climate change should be taken until scientific uncertainties are resolved, given that waiting to resolve all scientific uncertainties before action is taken may make it too late to prevent catastrophic human-induced climate change harms?
3. Do you deny that those who argue that they should be allowed to continue to emit greenhouse gases at levels that may be dangerous should assume the burden of proof that their actions are safe?
4. Do you deny that those who are most vulnerable to climate change's harshest potential impacts have a right to participate in a decision about whether to act to reduce the threat of climate change in the face of scientific uncertainty?
5. If you argue that the climate change impacts predicted by the IPCC have not reached a level of scientific certainty that warrants action, do you agree that climate change impacts predicted by the IPCC could be wrong in both directions and that impacts might be even harsher than those predicted?
6. If you acknowledge that human-induced climate change impacts could be harsher than those predicted by the IPCC,

do you deny that this possibility has ethical significance, including the creation of duties for high emitters to cease dangerous emissions levels?

7. Given that for almost two decades the United States has refused to commit to reduce its greenhouse gas emissions based upon the justification that there is too much scientific uncertainty to warrant action, if it turns out that human-induced climate change actually greatly harms the human health and environment of others, should the United States be responsible for the harms that could have been avoided if preventative action had been taken earlier?
8. Because climate change is a global problem, does any one nation have the right by itself to refuse to reduce the climate change threat based upon scientific uncertainty without giving those most vulnerable to climate change impacts the right to consent to be put at risk?
9. Because the longer the developed countries, including the United States, wait to reduce their greenhouse gas emissions on the basis of scientific uncertainty, the steeper the cuts needed to avoid dangerous climate change will be, if the mainstream view of climate change science proves to be correct, should the United States be expected to agree that it will be financially responsible for unavoidable climate change damages created by the delay if predicted climate change impacts are experienced?
10. Because one of the possibilities recognized by mainstream climate change science is that the earth could experience rapid nonlinear climate change impacts that outstrip the ability of some people and nations to adapt, should this fact affect who should have the burden of proof of determining whether climate change is safe or dangerous?

Section 3.

NANOTECHNOLOGY

13.

VALUE-SENSITIVE DESIGN AND NANOTECHNOLOGY

Ronald Sandler

Technologies are commonly understood as being complex tools that people use to help to satisfy their needs or wants. This "technology as tool" view is partially accurate. Information technologies, materials technologies, medical technologies, and agricultural technologies certainly help people to accomplish things they would not otherwise be capable of doing. However, technologies are not just tools. They are not just things that people use as means to ends. They shape our world, influence our perspectives, and mediate our relationships. They help to constitute our experiences and our power. They are crucial to our form of life. As a result, technologies are not value neutral. The implication of this for the process of creating technologies—engineering—is that it needs to be value sensitive. The first part of this chapter discusses the ways in which technology and engineering are value laden. The second part of the chapter discusses the concept of value-sensitive design. Although examples used in this chapter will be drawn primarily from nanoscale science and engineering, the conclusions regarding value and design apply to engineering and technologies more broadly.

TECHNOLOGY AND SOCIETY

Technology is a thoroughly social phenomenon. Some technologies are encouraged by society—e.g., by social demand or public funding. Other technologies are opposed or rejected by society (or at least some members of society)—e.g., genetically modified foods in parts of Europe. Technologies are always implemented in and disseminated through society. Sometimes they help to solve social problems—as many medical technologies do—and sometimes they create social problems—such as environmental degradation. In no cases are technologies separate from social context. They are all, always, socially situated. Every instance of technology creation and use is historical—i.e., it occurs in a particular place, time, and circumstance.

Technology shapes the spaces we inhabit. In our homes, businesses, and public buildings almost every aspect of our physical space is structured by technology. This has social implications. It influences who we see, how long we see them, and the conditions where those interactions occur. Technology also shapes broader spaces. Many cities and towns are organized in ways that accommodate themselves to car travel. Entire geographical areas—e.g., the Midwestern United States—have been transformed by technology. Where there were once vast prairies and woodlands, there are now vast farmlands. This impacts who lives (and visits) there, what they do, and how they interact with each other. Social interaction is influenced by the structure and types of places we inhabit, and the places we inhabit are shaped by technology.

Technology also shapes our conceptions of sociability—i.e., how we conceive of social life and what constitutes social relationships. Perhaps the clearest example of this is the impact of information technologies on social interactions. Cell phones, Skype™, Facebook® (and other social networking tools), massively multiplayer online gaming, and Second Life® (and other virtual realities) have opened up new forms of social interaction and new types of social relationships. As a result of these technologies, physical proximity is less and less a crucial component of meaningful social interaction. These technologies have also altered our social expectations. We expect to be able to reach our friends and family at almost any time, and plans are made in real time (as opposed to set much in advance). These technologies have, in a sense, extended space and compressed time with respect to social interaction, and they have altered how we spend time with people, and how many people we can spend time with. They have transformed our social worlds.

They have also transformed social institutions and organizations. Universities are vastly different today, as a result of information technologies, than they were twenty years ago. Students and faculty access and exchange information differently, they interact with each other differently, their education experience (both in the classroom and outside the classroom) is different. There are new parts (e.g., information services) and new priorities (e.g., online education and distance learning). How news agencies operate and information is delivered has changed dramatically due to information technologies. Government services and information are increasingly delivered online. The examples could go on and on.

As already mentioned, all of this changes our social expectations. We expect to access information quickly and easily. We expect to be able to reach people at any time. We expect government to make information available and to be transparent. We expect that people will be able to learn more about us (and to do so more quickly) than before. We expect to be able to rapidly travel large distances. We also expect to live longer, healthier, more comfortable lives than have people previously. In this way, medical technologies have impacted our conceptions of human flourishing. Life expectancy has increased over 40 percent in technologized nations over the past hundred years, and people expect to be largely healthy, comfortable, and active until the end of life.[1] Technology also mediates how we interact with the natural environment, from binoculars and birding, to nature programs on television, to coverage of environmental disasters (e.g., the BP oil spill in the Gulf of Mexico).

These are just some of the diverse ways in which technology impacts values and valuing.

Finally, technology shapes our daily lives and experiences. From the moment we wake up, we are dealing with technology. Many of us spend large amounts of our days looking at a monitor and punching buttons. Many of us take our recreation in ways that involve technology—e.g., television, video games, off-road vehicles, and electronic darts. As already mentioned, technology impacts how we interact with other people and the environment. All of us, almost all the time, are being impacted by technology in nontrivial ways.

TECHNOLOGY AND HUMAN BEINGS

That technology is so socially significant is an implication of the fact that it is fundamental to and inseparable from the human form of life. Our capacity as technological animals distinguishes us from all other species. Many species, such as starlings and dolphins, have complex communications systems. Many species, such as honey bees and meerkat, have elaborate social systems. Many species, such as elephants and octopi, exhibit social learning and tool use. However, no other species that we are aware of innovates, accumulates, and transmits ideas, information, and practices on the scale or at the rate that we do. A thermostat is a far more complex tool than anything found in the nonhuman world—let alone a smart phone or a space shuttle. A university is far more complex than any social structure found in nature—let alone a democratic state or the Catholic Church. Moreover, our social systems and tools change much more rapidly than does anything found in the nonhuman world. As already mentioned, the way universities function today—e.g., online libraries, smart classrooms, distance learning—is much different from how they functioned twenty years ago, prior to widespread personal computing, digitization of information, and the Internet. The social systems of wolves and the tool use of chimpanzees—two of the most psychologically complex nonhuman species—have changed comparatively very little in that time.

We, *Homo sapiens*, have a characteristic form of life, a characteristic way of going about the world, and that is the cultural way. We have the capacity, far greater than that of any other species, to imagine how the world might otherwise be, deliberate on whether we ought to try to bring those alternatives about, devise and attempt to implement strategies for realizing those alternatives we judge to be desirable, and disseminate them (through teaching and learning) if they prove to be successful. We are *Homo sapiens*, *Homo ethicus*, *Homo sociabilis*, and *Homo faber*. Our life-form is characterized by our comparatively large capacity for gathering information, social interactions, moral agency, and technology; we are the cultural animal.

Our life-form is only possible because of our capacity for culture. A human being alone, without social cooperation and without technology, would have difficulty surviving very long in any environment. It is due to our cultural capabilities that we are able to survive and flourish in so many diverse ways and across a wide range of places. We depend on culture (social cooperation and technology) in almost every aspect of our lives—e.g., food production, shelter, clothing, transportation, and recreation.

The basis for our comparatively large cultural capacity is our biology. The robust psychological and cognitive abilities that make culture possible arise from the size, morphology, and features of our brain, and we have the brains that we do (and not the brains of rattlesnakes or chickadees) because of our DNA. Indeed, the DNA that "codes" for brains like ours evolved, in part, because increasingly large capacity for culture was fitness enhancing under obtaining environmental conditions, and other parts of human biology (e.g., skull size and language capabilities) coevolved with the cultural capacities.[2]

Our cultural capacities are in these ways central to our form of life. Technology is among the most prominent components of culture. We are the engineering animal. Creating and using technology are quintessential human activities.

ENGINEERING, DESIGN, AND VALUE

Thus far this chapter has focused on the philosophical, cultural, and social significance of technology. It has been argued that the creation and use of technology is a quintessential human activity and the technology helps to configure every aspect of our social world. For these reasons, we need an ethic of technology.[3] We need deliberative tools (e.g., case studies, models, frameworks, and principles) and methods (e.g., deliberative approaches, forums, and educational resources) that can help us to reflect well on what technologies we ought to try to create or promote and how we ought to try to implement and disseminate them. The remainder of this chapter concerns the portion of ethics that concerns technology creation—i.e., engineering.

What software engineering, mechanical engineering, electrical engineering, civil engineering, and genetic engineering have in common in virtue of which they are each a variety of engineering is that they involve the creation and application of technology to address some problem or overcome some barrier. Engineers make use of scientific and mathematical knowledge to devise technological solutions. A core element of engineering is design—since it involves construction of parts, processes, and systems—and an ineliminable component of design is choice. Engineers must choose what problems they will work on; they must choose how they plan to approach the problems; they must choose what counts as an adequate solution to the problem. There is no engineering without design, and there is no design without choice.

Engineers are thus frequently confronted with choices about ends (or goals) and means. Such choices are value laden. To choose one end or goal over another is to make a judgment about the relative value of the goals, particularly when the choices involve allocation of scarce resources and have opportunity costs. Similarly, to choose a particular means (or method) over another is to make a judgment about the appropriateness of the means (i.e., that it does not violate any constraints), as well as its relative value (i.e., that it is more efficient or less hazardous) in comparison to other possible means (or methods). Given that engineering involves design, which involves choice, which involves value, and that the product of engineering is technology (which as we have seen is highly socially significant), good engineering practice requires technical and scientific expertise, as well as social and ethical awareness and responsiveness. Engineering is value sensitive when the full range of relevant values—economic, ecological, and social—inform design choices. Engineering in a value-sensitive way begins with identifying design choices and making explicit the values at stake.

The next section discusses nanoscale science and engineering and some of its applications in medicine. The subsequent sections discuss three core components of value-sensitive design: choosing a project; defining success for a project; and attending to means and byproducts. Many of the examples used in the discussion are from nanomedicine.

NANOMEDICINE

Nanomedicine refers to medical research and technology that makes use of nansocale science and engineering. A nanometer is one billionth of a meter, and the nanoscale is the scale of individual atoms and molecules. A gold atom is approximately .14 nm in diameter, a water molecule is approximately .25 nm in diameter, and DNA is about 2.5 nm in diameter. Thus, nanoscale science and engineering refers to the study of materials and systems at the atomic, molecular, and macromolecular scale, as well as the creation of technologies that make use of the properties of materials and features of systems at that scale.

Nanoscale science and engineering is a powerful technological platform for several reasons. First, materials have diverse and novel properties (e.g., electrical, chemical, and optical) on the nanoscale; properties that they do not have in bulk forms. There are several reasons for this. There are higher surface to volume and surface to mass ratios with nanomaterials than with their bulk counterparts, which often increases or alters reactivity; at the nanoscale both quantum and classical forces are significant (whereas in bulk form quantum forces are less significant); and materials take on novel forms or structures at the nanoscale. Consider, for example, carbon nanotubes. A carbon nanotube is a grapheme sheet (a lattice of carbon atoms) rolled up. However, different carbon nanotubes have different properties depending upon their diameter, whether they are single walled or multi-walled, and the direction that they are rolled (i.e., the orientation of the lattice relative to the role). And this is only carbon nanotubes and graphene. At the nanoscale, carbon also forms buckyballs (fullerenes), as well as other structures, and there are, of course, numerous other types of materials/particles besides carbon. Nanoscale engineering thereby provides engineers with more properties—and thus more possibilities—with which to engineer solutions to problems. In the area of nanomedicine, for example, this provides more possibilities for imaging agents and more possibilities for drug design.

Second, nanoscale science and engineering (again due to its scale) enables tight clustering of components and functionalities. Nanotechnologies can be engineered to do more things in less space than could technologies designed at larger scales. In the area of nanomedicine, this has led people to pursue *theranostics*, technologies that would diagnose for illness and deliver treatment (as well as evaluate the success of the treatments). It has also led to development of "lab on a chip" technologies that could perform multiple functions in one device. One example of this, in the area of nanomedical cancer detection, is the development of cancer detection arrays. Different cancers produce different biomarkers, distinctive molecules that are found in very small quantities in the bloodstream of people who have the cancer. Researchers have developed arrays of nanoscale cantilevers on which different antibody receptors are sited that are tuned to different biomarkers. When blood is dropped onto the array, if the biomarker for a cancer is present, it binds to the antibody receptor on a cantilever, thereby altering the conductance properties of the cantilever. Thus, it is possible to quickly and accurately test for a broad array of cancers (as many as there are cantilevers and receptors on the device), without lab machinery.[4] Some have even proposed that this could be an off the shelf diagnostic (as with a pregnancy test).[5]

Third, as illustrated by the case described above, nanoscale science and engineering enables greater precision and selectivity in engineering than has been the case previously. It is only because the antibodies can be tuned to particular biomarkers (and those biomarkers can be identified in the first place) and sited to individual cantilevers that the solution to the problem (early detection of cancer) is possible. Another example of this from nanomedicine is in targeted drug delivery. A limitation of chemotherapy treatments for cancer is that they are indiscriminate—i.e., they destroy healthy as well as cancerous cells. Treatment for cancer would be much improved if it were possible to more precisely target drugs and therapies to the diseased cells. Nanoscale science and engineering make this possible. First, it enables researchers to more finely and systematically image cellular systems, and thereby identify more precisely any distinctive features of cancer tumors. It is now known that the blood vessels that form around cancer cells (as well as other diseased tissues) are malformed, often having nanoscale perforations in them. Nanomedical researchers have attempted to use this to design therapy-delivery systems that slip through those perforations, thereby targeting the therapy on the tumor. In one instance, researchers engineered nanoscale gold shells with silicon cores that slip through the perforations and accumulate around the tumors. They are then irradiated with near infrared light, which heats the shells (because the gold absorbs the heat) to kill the tumors.[6]

Fourth, as mentioned above, nanoscale science and engineering provides tools for better understanding of engineering problems. In bioimaging, for example, quantum dots (nanoscale crystalline structures) are providing finer resolution imaging than was previously possible, and it is expected that networks of nanoscale biosensors will provide much more detailed and comprehensive data on system processes.

Nanoscale science and engineering promises to provide for better information regarding a broad range of engineering problems, more tools (properties) for attempting to devise solutions, greater precision in design, and more functionalities in less space. This is what makes nanoscale science and engineering a powerful technological platform in general (e.g., across engineering domains such as electronics, agriculture, materials), as well as with respect to medical technologies in particular.

VALUE-SENSITIVE DESIGN: CHOOSING A PROJECT

Because engineering involves trying to develop solutions to problems, choosing an engineering project involves choosing a problem on which to work (which often involves choosing a firm for which to work). There are a number of pragmatic considerations that shape substantially this choice—e.g., one's area of expertise, available facilities and resources, and institutional context. However, even given such factors, there is often a range of possible problems on which one might focus, and unless one chooses randomly there will be a basis for choice. That is, it will be an expression of one's values. To choose one end or goal over another is to make a judgment about the relative value of the goals, particularly when the choices involve allocation of scarce resources and opportunity costs.

People rarely choose goals that are despicable. Most people, most of the time, choose things because they believe they are valuable or desirable. Engineers are no exception. Except in rare cases—e.g., malicious viruses—the goal is to bring about some perceived good. With respect to nanotechnology, most people working in the field believe that nanoscale science and engineering in general, and their projects in particular, will be socially beneficial. After all, people living in industrialized nations today live healthier, longer, more secure, more comfortable lives than people at any other time in human history, and technology contributes enormously to this. Life expectancy in the United States for individuals born in 2005 is seventy-eight, whereas in 1900 it was forty-seven.[7] The vast majority of US citizens have reliable access to basic resources—water, sanitation, electricity, shelter, and food—and most have substantial additional economic resources.[8] If nanoscale science and engineering are the next powerful technology platform, then it seems likely it will help to make at least some people better off.

This would seem especially true of nanomedicine, since the point of nanomedicine is to take advantage of the characteristic features of the nanoscale to build more and better technologies for promoting human health. Certainly promoting human health is good. Therefore, nanomedicine is good. However, things are not as uncomplicated as they appear. First, no one does research on nanomedicine in general. All research is on some particular problem—e.g., diagnosis, treatment, information management—and/or on a particular disease—e.g., cancer or malaria. It also uses some particular technological approach. So there is no such thing as nanomedical research in abstract, just a lot of research projects using nanoscale science and technology on medical or health-related problems. So even within nanomedicine there is the issue of choosing which problem on which to work. This is a value-laden choice.

Suppose one chooses to try to develop an improved cancer detection technology. Such a technology would be good, since suffering and death from cancer is bad and early detection would likely help reduce these. However, that choice has opportunity costs. If one is working on cancer detection, then one is not using those resources to work on malaria prevention, vaccine storage in places with unreliable electricity, or durable and effective water purification methods. Such things are also good. It is arguably the case that the latter are greater goods, since they would have a larger public health impact (1.1 billion people in the world lack access to potable water). Moreover, they would help people who are less well off, since they are problems faced largely by the global poor—i.e., the 2.5 billion who live on less than $2 ppp/day, with 980 million on less than $1 ppp/day.[9] Since the large proportion of people who die from cancer do so late in life, whereas a large proportion of those who die from malaria do so very young, the per beneficiary longevity gains (the years of human life saved or suffering prevented) are likely to be greater with malaria treatments and water purification than with cancer detection and treatment.

In addition, if one is working on cancer detection, then one is not using those resources to work on cancer prevention. There is some evidence that there are significant environmental causes for increased rates of many types of cancers.[10] If this is correct, then there are two general ways to try to decrease death and suffering from these cancers. One can either try to reduce the incidence of cancer by reducing or eliminating the contributing environmental causes or one can try to improve detection and treatment. These are not mutually exclusive. However, if a particular research project is spending resources (machines, money, expertise, time) on developing better detections, then they are not using those resources to work on identifying and finding alternatives to the environmental causes. Again, cancer detection is good, but it might not be the better good. One reason might be that it is better to prevent something bad from happening in the first place than it is to try to eliminate a bad after it materializes. A second reason might be that if the triggers of cancer are reduced in the environment then there may be broadly distributed benefits, whereas novel technological detections and treatments are likely to be expensive and, particularly in countries where there is not universal health care, this means they will not be available to everyone (i.e., the benefits will be more restricted).

It is also crucial to consider whether novel, sophisticated technological approaches to resolving a problem are preferable to less technologically oriented approaches. For example, suppose a neighborhood has high incidence of childhood asthma. Certainly it would be good to reduce the number of severe asthma attacks. One way to do this might be to use nanoscale science and engineering to develop sensitive environmental monitors placed strategically in homes and throughout the neighborhood to continuously measure levels of environmental triggers (e.g., pollutants) and transmit that data in real time to people in the community, as well as to identify people in the community with any genetic factors associated with asthma in general or particular environmental triggers. Community members could then use this information to estimate the risk of an attack at a particular time and change their behavior (e.g., how active they are, whether they stay indoors, and whether they keep inhalers close at hand) as appropriate to try to avoid (or to decrease the likelihood) of an attack. This would be a preventative approach, one that if executed would be technologically impressive. However, another, much simpler, likely less expensive and more likely to succeed approach to prevention (because it depends on fewer variables) would be to identify the sources of the triggers in the community and, through policy, reduce them—by changing the location of bus depots, introducing and enforcing nonidling laws, eliminating incinerators, and so on. The benefits of such an approach would be broader based—everyone in the community would have cleaner air to breath—and there would be less that would have to go right to accomplish the goal of asthma reduction. Again, these approaches are not exclusive, but it would be inefficient to spend research dollars on problems that can be managed much more easily and less expensively in less technologically oriented ways. Moreover, if the promise of a technological solution leads people to not pursue the nontechnological approaches, then hyping the technological approach would appear to be a kind of negligence.

It is important to be clear about what is (and is not) being suggested. The idea is not that nanotechnology research on cancer detection and treatment is bad or that those who are expending resources on it are doing something wrong. What is being claimed is that there are opportunity costs with any research, and so when a choice to do something good or valuable is made that good is being prioritized over other goods. There are many possible reasons some problem might be prioritized—e.g., economic opportunity. But such choices are also expressions of values and priorities. So when making a decision on which project on which to work—even within as uncontroversial an area as nanomedicine—it is important that those values be made explicit and considered. And, crucially, choosing a problem in this case is not choosing a technology or technical approach with which to work. It is choosing a social or ecological problem to try to help address. Choosing, within nanomedicine, to use the science and technology platform to improve treatments for erectile dysfunction or develop methods of cognitive enhancement, rather than on reducing cancer deaths or purifying drinking water, is to express some values rather than others. The same is true if one chooses to work on military technologies or gaming technologies rather than nanomedicine.

The foregoing discussion suggests several questions one might consider when conducting a values analysis with respect to project or problem choice:

1. Who would (and who would not) be benefited or advantaged by solving this problem?
2. How large a benefit or advantage would it be?
3. Are there other problems that might be addressed using these resources, which would benefit more people, people who are worse off, or would confer a greater benefit?
4. Would this address the cause of the problem or would it treat the problem's effects?
5. Are there other, perhaps even less technologically sophisticated, approaches to addressing the problem that might be more effective, efficient, or likely to succeed?
6. Would anyone be disadvantaged or made worse off if this problem were solved?

One final comment on choosing projects: The more basic the research program—e.g., characterizing the basic properties of nanomaterials vs. building an application—the less available information there will be with which to try to answer these question. In some cases the research may be so basic that no reasonable or informative answer to these questions can be provided. But even in cases where the research project is applied (e.g., developing filtration technologies or energy technologies) there will, of necessity, be some speculation. This is a feature of all proactive ethics. However, it is particularly prominent in the case of value-sensitive design, and it gives rise to what is sometimes called the Collingridge dilemma: on the one hand, the earlier in the technology development process that values are considered, the less information we will have about the technology and its impacts with which to work; but on the other hand, the later in the technology development process that values are considered, the less opportunity there is to impact the trajectory of the development since the research path will be more defined.[11]

Because value-sensitive design is proactive, it must proceed without ideal levels of knowledge. As the discussion in this section indicates, this does not mean that reasonable and well-informed answers to the above questions are not possible. But it does mean that one must be sensitive to the limits of one's own knowledge, and that to address the information deficit one must make use of all available resources—e.g., historical analogs, case studies, experiences of those who have dealt with similar problems in the past, and relevant facts about the social and ecological state of the world. One cannot focus just on the features of the problem or the technologies themselves. There is a difference between uninformed and informed speculations, as well as between wild and measured speculations. Value-sensitive design involves trying to make measured, well-informed speculations about the likely impacts of different research programs.

VALUE-SENSITIVE DESIGN: DEFINING SUCCESS

Closely related to the issue of choosing a problem on which to work is defining success with respect to the problem. Again, it is crucial to distinguish the technology that (it is hoped) would bring about the success from the success itself. Creating a nanotechnology-based cancer detection system is not solving the problem, reducing suffering and death from cancer is. So even if the technology works in the lab, if it does not do so in a way that can be translated outside the lab (e.g., due to technical constraints or costs) it is not a successful technology.

A clear illustration, from outside nanomedicine, of value sensitivity in defining technological success and how that translates into design choices is the case of the $100 laptop or the One Laptop Per Child initiative. The initiative has the proximate goal "to empower the children of developing countries to learn" and to "create educational opportunities for the world's poorest children."[12] The ultimate goal of this is to improve the quality of life for these children and their families. The means for achieving the goal is to increase information and creative tools to children, and the means to deliver that is laptops.

Given this definition of success, producing a $100 laptop is not sufficient for the project to meet its goal. The project would only be successful if the laptops were affordable, functioned well in the conditions in which the children lived (e.g., unreliable or no access to electricity, unreliable or no Internet provider, and little if any technology support), were disseminated, and were used by students to create and learn. The laptops needed to be designed in ways that were conducive to accomplishing these. For example, its screen is readable in bright sunlight, since many children's classrooms are outdoors. Its keyboard has a sealed, rubber membrane to protect it from humidity. It can be charged on alternate power sources, such as car batteries. It has no hard drive and only two internal cables. Its software is open source. It has a long-range antenna. All aspects of the machine were designed with sensitivity to what was defined as success for the project.

The project, thus far, has had only limited success. Just 2.5 million machines have been made, and because the volume has been lower than expected, cost per machine has not dropped below $180.[13] This means that the laptops are not getting in the hands of as many children as anticipated, and the educational and social targets are not being met. The program has not been a complete failure. Some machines are in the hands of children; it has spurred other initiatives for low-cost, high-access laptops; some of the components have been adapted to other purposes; and lessons were learned for future initiatives (including the $75 One Tablet Per Child program). Still, it has not had the transformative educational effect that was the defined target for success.[14]

Nevertheless, the One Laptop effort illustrates how robustly a conception of what constitutes success can permeate technology design (as well as distribution strategies). Value-sensitive design therefore requires clarity on what is the standard for success, and sensitivity on how design choices are likely to impact whether success is achieved. Consider, again, nanomedical cancer detection and treatment. One might defines success in terms of lab or trials accomplishments, making it to the marketplace, making money, saving lives (or producing healthy years), increasing public health, or benefiting the worst off. If the goal is to be widely disseminated, then the capacity for low-cost mass production will inform design in a way that it would not if the goal is proof in concept. If success is defined in terms of revenue, rather than in terms of benefit to the global poor, then one is not likely to use open source technologies.

The above discussion suggests several questions that would be useful to consider in defining success for a project:

1. What outcomes would constitute success for this project (in relation to the problem being addressed)?
2. What is technical success for the project and how does it differ from social or ecological success?
3. How is success to be measured, or outcomes evaluated—i.e., how does one come to know the extent of one's success?
4. Which aspects of technology design could be altered to increase success as it is defined?
5. What social factors are relevant and how can they be addressed through design?

One final comment on defining success: The social dimensions of technology are highly relevant to whether success is achieved. Technology by itself never solves a problem. For example, technologies that could be beneficial to the global poor often never get deployed (even if they are created) because they are not in a form well-fitted to people's needs, living conditions, or culture, or they are not manufactured and disseminated due to policy, infrastructure, or cost constraints. How the technologies are deployed is also crucial. The fact that nanotechnologies are being developed that increase available supplies of useful or potable water, for example, does not ensure that those technologies will be social or environmental goods—i.e., that they will be successful. If they are deployed in ways that enable cultivation of water-intensive crops in arid locations or encourage population migrations to unsustainable locations, they may perpetuate and create problems, rather than resolve them. This is to point out the depth of the challenge and the limits of engineering. As exemplified by the One Laptop project, even when one chooses a highly significant problem, clearly and reflectively defines success, and has that definition robustly inform engineering decisions, the project may well not fully succeed (and that was a comparatively very well-funded project). The reasons for this might have something to do with the technology (e.g., the cost structure or reliability), but it might also have to do with social factors (e.g., perceived value, alternatives, or trade policies). Value-sensitive design requires attending to both the technical and social factors relevant to the success of a technology.

VALUE-SENSITIVE DESIGN: MEANS AND BYPRODUCTS

A third dimension of value-sensitive design, one that is often closely related to defining success, has to do with reflecting on the means by which success is achieved and the byproducts of the pursuit of success. As is commonly recognized, some means are not ethically acceptable, even if their ends are worthwhile—this is why there is regulation of medical research involving human subjects, for example. Good intentions and a laudable goal are not sufficient to ensure ethically acceptable practice. Moreover, there may be unintended outcomes of technology development and dissemination that need to be avoided. In these ways, sensitivity to value requires sensitivity to how technology is developed and to secondary or unintended outcomes of technology development.

This type of sensitivity to value is implicit in attempts to reduce the ecological impacts of products. Life cycle assessment (LCA), for example, is a tool that can identify the ecological costs and benefits of a product along the entire production, use, and disposal process. LCA is not a decision tool—it does not tell one what to do. But it does provide information that can substantially inform design decisions. If one is committed to the value of reducing the ecological impact of a technology, LCA can be used to determine where in the lifecycle of a technology (or process) particular ecological impacts occur and can help to identify less impactful alternatives. The use of LCA to inform engineering designs in order to prevent an undesirable byproduct is a paradigmatic and increasingly common example of value-sensitive design. It is sensitive to the value of ecological and human health.

Again, such sensitivity is crucial even in cases where the primary goal, the problem to be solved by the technology, is not problematic. For example, silver nanoparticles, which are an antimicrobial, have been introduced into some washing machines to try to better remove odor-causing bacteria from clothes. One concern about this is that the particles will be released into the sewer system and will have an adverse effect on water treatment, which makes use of microbials. This would be an unintended and negative byproduct of the technology. Value sensitive design requires either developing the technology in a way that prevents the release of the particles, or evaluation of the release to determine whether there is indeed a potential problem. The addition of silver nanoparticles to washing machines is not such a great good that it justified taking on significant risks.

Another example, again from nanotechnology, concerns privacy. There are concerns that nanoscale sensor technologies could pose challenges to personal privacy due to their scale, potentially low cost, and multiple functionalities—e.g., data collection, analysis, and transmission. There is some precedent for this with the case of radio frequency identification tags (RFID). These tags are increasingly embedded in objects that we use and carry around in different contexts. They can provide information about our whereabouts (e.g., toll road passes and transit cards) and our tastes (e.g., clothes we pull off racks in stores), for example. They can also be hacked. As these technologies continue to permeate the products that we use and have increasingly personal information encoded in them (e.g., medical records) there are increased concerns about the amount of data that will be available about us, how it will be used, and who will have access to it.[15] This has led some researchers to suggest that nanoscale biosensors and information technologies need to be engineered in ways that decrease the likelihood of information about ourselves becoming available to those whom we do not want to have access to it[16]—e.g., designed so that they can be more easily detected, less easily hacked, and transmit information more precisely. These may not be crucial to the primary function of the technologies, but they could be helpful in preventing undesirable secondary effects.

What these examples illustrate is that there are a range issues, not related to the primary goal of the technology, which can function as constraints on the development of the technology. Technology design is always conducted under constraints—most obvious of these are cost and material availability. But these cases show that there can be appropriate value constraints as well—e.g., grounded in environmental values and individual liberties. Value-sensitive design needs to take these into account by asking:

1. What are some potential unintended byproducts of the development, production, and use of this technology?
2. Would any of those byproducts be problematic?
3. How problematic would they be?
4. Could those unintended, undesirable products be avoided or mitigated through design of the technology?

As with the other aspects of values-sensitive design, there is a projective element here. However, one cannot merely speculate or rely on old dogmas. Value-sensitive design takes considerable commitment and effort. As the examples of LCA and RFID indicate, there are tools and experiences with which to work. In addition, engineers can collaborate with social scientist and science, technology, and society (STS) researchers who specialize in doing technology assessment and who study the social and ecological impacts of technology.[17]

CONCLUSION

Engineering is a thoroughly value-laden enterprise. Value-sensitive design involves identifying value judgments in engineering design processes—e.g., choosing projects, defining success, delineating constraints, and minimizing problematic secondary effects. It also involves cultivating space within the engineering process for reflective discourse on those judgments and how they can and should inform design decisions. In addition, it includes developing the capacity for productive reflective discourse—e.g., strategies to help identify the full range of alternatives and evaluative frameworks that encourage attending to all relevant considerations. Even in a case where the goal (e.g., health and longevity) is widely regarded as good and desirable, nontrivial value-laden choices abound. This chapter has tried to illustrate core components of value-sensitive design, as well as suggest some key questions that when posed in engineering and technological innovation contexts can serve to encourage the value-sensitive design process.

14.

DEBATING NANOETHICS

US Public Perceptions of Nanotech Applications for Energy and Environment

Barbara Herr Harthorn, Jennifer Rogers, Christine Shearer and Tyronne Martin

"Nanoethics" encompasses a series of debates about the nature and specificity of moral concerns raised by nanotechnologies.[1] In this chapter we present and discuss the US public's ethical/moral reasoning and cultural values regarding nanotechnology research and development (R&D) in the context of deliberative workshops on nanotech applications for energy and environment. The work is drawn from a series of deliberative workshops conducted with US public participants in California in 2009, with additional insights from a series of cross-national deliberations conducted in 2007 in the United States and the United Kingdom. Both within the US context and US–UK comparative study, our work has been directed not on producing consensus views, but on highlighting the diversity of views about emerging technologies held by multiple parties and multiple cultures. We argue that understanding and addressing the multivocality of public views is essential to deliberative democracy generally, and to the US national project of "regulating risk and ethical concerns"[2] of nanotechnology R&D more specifically.

One thing that is relatively new in the nanotech context is the extent of US governmental concern about ethical, legal, and societal issues (ELSI)[3], evident in the small but consistent funding for this purpose through the National Science Foundation as a part of the National Nanotechnology Initiative (NNI).[4] Although engineered nanomaterials (ENMs) do display novel physical, chemical, and optical properties, are the ethical concerns raised by their development and implementation different from those of other technologies? There are conflicting views on this—historians of science emphasize the recurrence of past modes in nanotech scientific developments,[5] and some physical scientists argue nanomaterials are no different in principle from any other synthetic chemicals.[6] Yet the interdisciplinarity of the NSE enterprise is argued by some to exceed that of past scientific developments with new scientific and social effects, and the environmental, health, and safety issues in particular for this very large class of materials pose an unprecedented scope of possible risk assessments, and thus, potentially, problems that are different in kind as well.[7] Is the innovation system for nano different from past R&D in ways that are more equitably distributed? It is indeed a highly dispersed technological development system being built through extensive international collaborations as well as intense international competition, but it is not yet clear if global nano development will on the whole contribute to accelerating global economic inequality or serve to ameliorate some of these inequities.[8] And on the public side, upstream perception research has thus far primarily shown benefit-centric views and a positive link between knowledge about nano and perceived benefits.[9] The latter is an unusual finding in the psychometric risk perception and public attitude fields, but it is not possible to say yet if it derives from distinctive public views of nanotechnologies or simply from the unusual upstream, prerisk event context in which the views are being elicited and measured. The challenges of upstream ELSI research carry into the deliberative context as well, as participants are asked to form ideas and judgments about new technologies whose very existence may have been unfamiliar to them before they walked in the door.[10]

"Responsible development" of nanotechnology concerns a multitude of factors that are described as contributing to (1) the commercialization of widely beneficial applications, (2) simultaneous planning for societal changes that could result from such innovation, and (3) the avoidance of potentially negative impacts from new products. "In the National Research Council committee's view, responsible development of nanotechnology can be characterized as the balancing of efforts to maximize the technology's positive contributions and minimize its negative consequences. Thus, responsible development involves an examination both of applications and of potential implications. It implies a commitment to develop and use technology to help meet the most pressing human and societal needs, while making every reasonable effort to anticipate and mitigate adverse implications or unintended consequences."[11] This chapter focuses instead on the social and cultural aspects of responsible development, which are notably absent from the above list. Addressing them involves a set of assumptions and practices for which we argue there are not yet any established templates. This does not necessarily mean that nano ethical challenges are unique or new. It does highlight, however, the lack of preparedness to connect political aims of responsible development with practical societal implementation.

Who should decide what constitutes "the most pressing human and societal needs," and through what process? Surprisingly, a recent article in the *Economist* (December 19, 2009) advocated a radical shift in the criteria used to assess technological development from the "impoverished" current view of "material progress" to what they termed "moral progress."[12] Drawing from the philosophical work of Susan Neiman in her book, *Moral Clarity*, this approach argues for a kind of social sustainability, where at key nodes in the process, we make decisions based on what is good for society rather than what will most enhance "material progress," and that we support those decisions through the legal and social system, using both incentives (e.g., tax breaks) and deterrents (e.g., prosecution).[13] They note that science (and of course industry) would need governing for such a shift to succeed. This appealing formulation of an enlightened government and industry depends, however, on a number of challenging, if not impossible, conditions.

The NNI was formulated around a similar set of aims: "Progress is faster with proper vision and choices guided by moral values, transformative goals, collective benefits, and professional ethics."[14] Yet how do we determine "what's good for society" or "proper vision and choices"? Assuming a completely dedicated governance system, whose views should matter in making such a determination? In a large, multicultural nation like the United States, will views coalesce about what constitutes "moral progress"? And if not, how do we devise a process for incorporating such multiple viewpoints?

Deliberative democracy is one answer to the question of how such potentially diverse values, goals, and views of risks and benefits can be engaged.[15] If government is invested in engaging in ethical regulation and development of nanotechnology, deliberative democracy is a way to disseminate scientific and technological information and foster an informed citizenry, to enhance public participation and "deliberative habits"[16] and, as we used it, serve as a method for studying public perceptions of nanotechnology. "Instead of trying to win a philosophical argument concerning the viability of first-order principles (e.g., efficiency, safety), deliberative democrats are more concerned with determining what would constitute a *reasonable balance* between conflicting fundamental values."[17] In addition, environmental decision analysts have contributed important methods and theory for addressing difficult tradeoffs in preferences and values in managing environmental decisions at the community level.[18]

We are interested in the shared (or not so shared) cultural values that surround social and ethical issues of nanotechnology. People carry with them perspectives and values that are associated with their lived experiences, gender, ethnicity, education, age, and occupation when they attend a deliberation.[19] In a discussion of nanoethics, this approach works to acknowledge and accommodate difference such that all voices can be heard. Therefore, we ask what are the key social and ethical issues associated with nanotechnology that emerge in our diversity-oriented public deliberative workshops?

SOME KEY ISSUES FOR ETHICAL DEVELOPMENT OF NANOTECHNOLOGIES

Governance and Trust. The resolution of governance issues and decision making about nanotech R&D is a pivotal issue. Larger issues of regulatory change to accommodate nanotech complexities are mired in political and institutional impediments that will be difficult to surmount. Yet doing so would appear to be essential for maintaining public trust in the R&D effort. Indeed, in our workshops, the need to regulate nanomaterials industries for the good of society was a strongly expressed normative value. Governance is seen as essential to limit risks to privacy, to ensure equitable distribution of benefits and risks, and to safeguard the nano workplace.[20] Potential mismanagement of risks, what sociologist William Freudenburg has called "recreancy," by government and industry is also seen as a serious constraint on public trust and the ability to realize the benefits of technologies in general,[21] nuclear technology,[22] or new nanotechnologies in particular.[23] Thus *trust* is an important thematic concern with strong ethical dimensions. Deliberative democracy is one mode for increasing public participation, public involvement, public dialogue and, presumably, public investment in the shared project of development (or decision making about it).

Justice and Equity. Justice is another critical ethical dimension in public deliberation on nanotech. There are a number of elements to values concerned with justice. Will the risks and benefits be equitably distributed within and between nations?[24] People in our work express diverging views about whether they will benefit from new nanotechnologies and, conversely, whether they will actually be harmed by known or unknown risks or unanticipated negative consequences. Environmental (in)justice is strongly implicated in amplified perception of risk.[25] More generally, even many of those who personally feel privileged regarding the likelihood of receiving benefits express strong ethical values toward more equitable distribution, including globally in the developing world. In our work, these distributional justice issues are important to the way people make sense of new technologies. Justice has another component as well in these discussions. Procedural justice—the fair treatment of all involved stakeholders—is another theme that emerges throughout.[26] In the upstream context where a large percentage of the population has little or no awareness of nanotechnology,[27] and hence the public has neither been informed about the risks of nanotechnologies nor consulted about exposure to potential risks, informed consent is a doubly critical justice issue.

Responsibility and Power. Who is responsible for nanotech "responsible development"? What is the scope of responsibility—community, state, nation, or world? In the context of energy and the environment, will people take responsibility for the commons? Whose responsibility is it to conserve and protect the environment? Who will pay for R&D? What are cultural values about the environment, and are there enough incentives to use alternative and renewable energy sources, given the current cultural and economic context? How will cost and access issues affect restrictions on adaptation of new technologies for energy and the environment? Paul Slovic has argued persuasively for several decades that the assessment of risk is *always* an exercise of power.[28]

Uncertainty and Precaution. As in all application domains of nanotech, uncertainty about physical hazards and social disruption is a recurrent issue. The ethical dimensions are clearly evident when uncertainty about risks is used as a basis for industry and government resistance to regulatory intervention. Uncertainty is called up as a rationale for unfettered R&D by prodevelopment members of the public (who often happen to be middle class or elite, younger, white men[29]) just as precaution is being advocated by the European Unon (REACH), nongovernmental organizations such as ETC, FOE–Australia, Greenpeace, and US occupational health and safety researchers.[30] Will nanotechnologies for alternative energy sources and environmental remediation pose new problems for the environment?

We cannot know the answers to these questions at this time, but research into public expectations, values, beliefs, and perceptions concerning these issues opens a window into the ethical and cultural values that pertain to the development of nanotechnologies and that mediate the possibilities for achieving responsible development.

METHODS AND APPROACH

The research reported here was designed to explore and develop new understandings of public expectations, values, beliefs, and perceptions regarding new nanotechnologies through qualitative research in deliberative public settings. In July through October 2009, our interdisciplinary research team of social and physical scientists ran one pilot and six public deliberation workshops in California. These workshops built on the 2007 Center for Nanotechnology in Society at the University of California, Santa Barbara, cross-national study's protocol for deliberative engagement about nanotechnologies, while adding a new emphasis on gender.[31] The half-day deliberative workshops focused on either energy and environment or health and human enhancement applications of nanotechnology, with gender composition varied systematically between the groups. Each workshop consisted of a focus-group size meeting in which we sequenced initial open discussion of the cultural contexts for thinking about energy/environment or health/enhancement, followed by carefully framed formal educational presentations. This was followed by an open reading time, in which participants chose from a large range of fairly brief written materials on nanotechnologies in general and in the relevant application domain and their potential implications. The group then broke into three smaller World Café tables for facilitated in-depth discussions. The workshops concluded with a final hour-long dialogue by the full group. Pre- and postmeasures were used to provide a limited set of scaled judgments and more open-ended comments. Sessions were audio- and videotaped, and full transcriptions of all dialogue were prepared based on these. Data analysis is primarily qualitative via formal content and narrative analysis using NVivo® software and systematic interpretive analyses by the team to determine key narrative themes and group dynamics.

We also draw on quantitative analysis of the pre- and posttests administered to all workshop participants.

Workshop content in general and examples of specific nanotechnology applications were vetted with Nanoscale Science and Engineering (NSE) collaborators and the team NSE graduate researcher for scientific accuracy and validity. In the energy and environment applications sessions, examples focused on technologies for energy conservation (e.g., energy efficient lighting), renewable energy (e.g., new stronger, lighter blades for wind generation; quantum dots for solar cells), and environmental remediation (e.g., sensors for detecting soil contamination; water filtration membranes).

Out of 67 total participants in six workshops, 33 were men and 34 were women, and each workshop ranged from 9–13 participants. With our aim of democratic inclusiveness, we recruited a diverse quasi-representative quota sample of participants designed to match as closely as possible the demographics of the central coast California area along race/ethnicity, occupation, age, and income criteria, and drawn from as diverse a set of recruitment points as possible. Following other deliberative research,[32] we held minimal education level to completion of high school.[33] A slight majority of the participants were White, 37 (55 percent), 15 (22 percent) were Latino, 6 (9 percent) were African American, 5 (7.5 percent) were Asian American, and 4 (6 percent) classified themselves as "Other." All participants were 18 or over. A slight majority (55 percent) had a bachelor's degree or higher, 34 percent completed some college or an associate's degree, and 10 percent had only a high school degree. Age was fairly evenly distributed: 16 percent were 18–25 years old, 13 percent were 26–35, 15 percent were 36–45, 12 percent were 46–55, 21 percent were 56–65, and about 22 percent were 66 or older. A majority (58 percent) of participants reported a family income under $50,000.

ETHICS AND VALUES IN DELIBERATIVE DISCUSSION

In the following we provide analysis of the key themes as expressed through workshop narratives to give a sense of the range and diversity of views, points on which even such diverse groups can achieve some consensus, and implications of these views for responsible development of nanotechnologies for energy and environment.

Governance and Trust

A critical aspect of governance is the decision making process, particularly regarding the management of risks and, though less frequently addressed, dissemination of benefits. Who makes decisions about new technologies (the "risk makers")?[34] Who decides which risks matter and which risks (and exposed bodies and environments) do not?[35] Who sets "safe exposure" guidelines, and what criteria do they use? In the US democratic context, of necessity this process involves a somewhat participatory approach that makes contentious debate likely[36] but also creates the possibility for ways forward in pluralistic decision making.[37] Participants expressed frustration at decisions being made about nano without their knowledge or input:

> "Well, I guess that is part of the fright for me in all of this, is the idea that this is not being done by the CDC or the Surgeon General Department. It's being developed and experimented with and promulgated by private companies." (Laetitia, Eng/Env #3, Female, African American, 66+)[38]

> "Maybe we should talk about ethically, who are they [academic scientists and technology developers] to decide what is right for society?" (Sal, Eng/Env #2, Male, East Indian, 18–25)

> "I just really feel that the things that are even currently out on the market, that I just learned about today, should definitely be approved by the FDA." (Nicole, Eng/Env #1, Female, White, 46–55)

Workshop participants generally supported risk research and regulation of nanotechnologies. This was circumvented, however, by a sense of urgency for particular energy and environmental applications, such as those that will offer alternative energy sources and relieve dependence on oil. A sense of urgency reduced perceived risk, or at least increased the perceived benefit high enough to outweigh any risks. As in the 2007 US–UK workshops,[39] perceived urgency appears to be a very strong driver of views about energy and environmental applications. Many participants shared the view that continued loss of natural resources and desperate need for new, affordable energy sources will drive motivation for developing new technologies as well as the adoption of alternative energy and conservation strategies.

> "If we do not develop new sources of energy, we're doomed. If we don't develop the technology to fix the environment that we've already ruined, it will just keep on." (Sal, Eng/Env #2, Male, East Indian, 18–25)

> "Maybe if we reach like a tipping point where there is no . . . we are just killing the planet so much that we have no other choice, like the necessity is the mother of invention. That we are just going to have to, we are going to have to make some type of renewable energy to keep us going because we have no other choice." (Carlos, Eng/Env #1, Male, Latino, 18–25)

> "I believe fuel is the most high point in everybody's forefront of their mind because they don't want to pay more money. We are going to run out of fuel you know, just things like that. Anything that is happening, affecting us right now is what is going to [be] most important to people because that is what is in their mind." (Cristina, Eng/Env #1, Female, Asian, 26–35)

As these quotes show, urgency works in a certain degree of opposition to "responsible development" or "moral progress" as a reasoned, thoughtful selection of the higher moral road to collective betterment. The implication is that desperation ("killing the planet" "environment that we've already ruined") is what ultimately forces people to do what is needed. The view of humans (or Americans) as crass, shortsighted, and motivated only by cost has a cynical edge to it, mirroring analyses of UK respondents to nanotechnologies in deliberative settings.[40]

Near the end of the workshop, we discussed whether the US public knows enough to participate in deliberation. The views, at the end of almost five hours of intensive discussion where participants contributed their views and intelligence on a wide range of social issues, were that "the public" was largely seen as ignorant and too limited in knowledge to share in decision making. Many of them expressed regret about this, but very few shared our view of them all as enabled social actors with views, ideas, concerns, and beliefs, as well as knowledge, that matters. In this paradoxical situation, a majority of them expressed enjoyment at participation, asked for resources to continue their learning process, were enthusiastic about being involved in future such events, including desiring to be added to our center's mailing list for public events announcements, yet they did not, in general, think they specifically or the wider public more generally were ready and able to contribute in formal (as opposed to experimental) deliberative or public participation mechanisms for decision making about these technologies.

> "And I trust the public knows that they are not competent on most of these major issues and that's why they elect governments to deliberate for them. Unfortunately, our governments are mostly incompetent right now, so I don't know who's going to do the work." (Bert, Eng/Env #2, Male, White, 66+)

> "No, I do not think the public [knows enough to participate]. I mean, I think the researchers and scientists at MIT, those are the people with the knowledge. You have to have the knowledge, the working knowledge, I think, to make, to make important decisions like that." (Lacy, Eng/Env #3, Female, White, 46–55)

> "Yeah, I don't think I know enough [to participate], and I don't think the general society just in the United States knows enough, let alone the world." (Lashawn, Eng/Env #3, Female, African American, 56–65)

Gastil has discussed the need for developing the "deliberative habits" of the American public.[41] Our work certainly highlights the current gap between the aims of achieving public participation in the national project of responsible development and the feelings of efficacy and preparedness on the part of many public citizens to engage in this way (and this was arguably a relatively confident and interested group of them). This need for enhanced deliberative capacity seems to us a vital issue to address if this goal is to be realized, particularly since so many participants expressed uneasiness over research and regulatory decisions being made without their input.

Justice and Equity

Equitable access for Americans seems to revolve first and foremost around cost issues. In these workshops and the 2007 study,[42] US participants (in contrast to UK participants in the 2007 study) tended to construe new nanotechnologies in a highly consumerist framework, in the mode of luxury consumer products. Their responses to questions about perceived benefits are colored, then, primarily by concerns about prohibitive cost (and lack of economic means) as a distributive justice issue. There is concern that only the rich will benefit and have access to the new technologies. On the other hand, economic inequality is also discussed as a disincentive to adopting alternative energy sources. Expectations of inequitability were often normalized among individuals who are struggling for economic survival or have faced other forms of discrimination:

> "So it is possible but it seems that the average person is too worried about getting laid off, feeding their family, they are not going to say, 'Oh, I am going to dedicate my year, and all of my savings and more to going on the solar powered energy.' They are not thinking about that right now." (Katie, Eng/Env #1, Female, White, 18–25)

> "I am unemployed. I have been unemployed since November, when I got laid off. I'm barely making ends meet. You know? And so the more we progress into this high tech world where some people are just going to get left out. . . . I don't know what it is going to do to our society." (Nicole, Eng/Env #1, Female, White, 45–55)

"It's gonna get better, it's just, it's gonna be unequal in the way it makes things better." (Lance, Health/HE #1, Male, White, 56–65)

Will those who would most benefit from nanotech energy and environment technologies get access?

"It [water purification technology] can benefit people a lot, but then again we talk about the costs. Who is going to pay for this? Are we going to pay for this [global development] as American tax payers to benefit. . . . I am tired of seeing my tax dollars be thrown into research to benefit someone else." (Jess, Eng/Env #2, Male, White, 36–45)

"I believe [clean] water for everyone is a moral imperative." (Randy, Eng/Env #2, Male, White, 26–35)

In general, people were alarmed to learn that nanotechnologies were already being disseminated in the absence of safety knowledge, but thought that more risk research, rather than a precautionary approach to handling engineered nanomaterials in labs, a moratorium, or other interventions, was the right solution to the problem.

"Lack of informed consent. That means, where do I get to choose whether or not that is in the environment? I mean the article I read had a real mention of . . . there can be nanoparticles floating in the air or in the water or in the food chain." (Jay, Health/HE #3, Male, White, 46–55). "Already existing or no?" (TM, facilitator). "Yeah, it could, already existing, could escape and then those can have unseen, unknown environmental impacts. You know, you think of like the cane toad in Australia where they brought that in to, you know, for natural pesticides, and then that thing just took over and they had no natural predators. I mean, how many, how many nano equivalents to that, you know?" (Jay).

"Sure these are new technologies [that] have got fabulous possibilities, but it's not fair or reasonable to just hoist them on us without our knowing and having the opportunity to say yes or no. I don't want that as part of my world." (Laetitia, Eng/Env #3, Female, African American, 66+)

> "It is already out there, we're not being told about it. We're consuming it. They don't know what the long-term effects are going to be. And it's already being put upon us. We're buying it. We're using it and we don't even know it." (Lashawn, Eng/Env #3, Female, African American, 56–65)

Responsibility and Power

Even though the urgent need for affordable, abundant energy was widely represented among these deliberative group participants, and largely untempered by any expressed values for conservation, many also thought that, in spite of urgency, economic incentives were needed for both industry and consumers to adopt new technologies. Many participants saw American people and companies as unmotivated on purely ethical grounds:

> "If there's incentives for it, like discounts or, you know, a promotion where the first month or two where it's cheaper to buy the green products and to use them and so that people see how it works and understand it more. Then it might be a better incentive for people to start getting the ball rolling and thinking in that direction. As opposed to just saying, 'Green is better, try this.'" (Cristina, Eng/Env #1, Female, Asian 26–35)

> "It seems there ought to be more incentives to drive our personal behavior and personal responsibility and I'm not talking about a tax. I'm talking about you earn something for good behavior." (Bob, Eng/Env #2, Male, White, 56–65)

In terms of the equitable development and consumption of nanotechnology, most of the group felt that while we all—as a society and as individuals—*should* be responsible, we do not all share the responsibility equally. Discussion of responsibility ranged in scope from the individual through community, nation, industry, and the globe; each entity holds some responsibility, but each fails to live up to that responsibility. In discussion of the role of industry and government, sentiments differed between the need for oversight and the use of incentives to encourage responsibility, against only a few participants who believed that companies are capable of self-regulation.

> "I am thinking more about assessing the risk on a worldwide basis as well. So I think the international body has to come together and say, okay, what is our objective here with this new technology, how are we going to govern it. How are we going to assess the risk and then distribute it?" (Ralph, Eng/Env #1, Male, Latino, 46–55)

> "I just want to jump in and say corporations [should be responsible], sure. I mean, they are going to be the ones creating these products, but they have to be governed, by the government. . . . [B]ecause you get people in there who are not morally right, they don't see the smaller people, they see the dollar and they have no accountability. They have to be accountable." (Nicole, Eng/Env #1, Female, White, 46–55)

> "I just believe that, in my opinion, I think the government is exactly like a corporation. It's just bigger and it has more responsibility over a larger amount of people. The more money that the government gets it is the better for them, the more money the corporate gets, it's better for them." (Cristina, Eng/Env#1, Female, Asian 26–35)

Uncertainty and Precaution

Discussion about the physical (EH&S) risks of nanomaterials and products necessarily involves a great deal of uncertainty in the current context.[43] Scientific uncertainty in the present, especially about the longer term risks, can amplify concerns.[44] Uncertainty over potential risks is also likely to be gendered and raced and hence to intersect issues of inequality and justice, as we would expect concerns about who will bear the risks to be filtered at least in part through past experiences of discrimination and vulnerability.[45]

> Cristina: "You can only do so much in a lab, too. You can test for everything that is conceivable, but there are just so many possibilities and factors out there that not, you can't get anything complete. . . ."

Nicole: "And study it for ten years."
Katie: "Yeah the ethical concern."
Nicole: "Study it for decades."
Cristina: "You could study it for like 50, 100, years and you still wouldn't get every possibility of everything." (September 19)

On the issue of nanoscale cerium oxide additives in fuel:

"So often when there are negative effects of anything in the environment on children—the same negative effects can be applied to many people with emphysema, breathing difficulties, asthma, and the elderly. And so it's not just we are going to sacrifice this, you know, beautiful little three-year-old, we are going to sacrifice a beautiful little three-year-old and half his family, up to grandma." (Laetitia, Eng/Env #3, Female, African American, 66+)

"And we do not know what the long term consequences look like. . . . If they [at-risk children] start being damaged at three years old because we think it's more important that we whiz around, whiz around, whiz around." (Lashawn, Eng/Env #3, Female, African American, 56–65)

"And unfortunately that is a lot of stuff that happens with development of new technology is you can't always know what happens with [the] development of new technology. You can't always know what is going to happen with all the scenarios and factors that come into play. And it's a moral debate. Do we, do you use it now for the benefit of saving a thousand people because they didn't have enough good drinking water? They are drinking muddy water from, you know wells and what not, and have those people survived, right now? Or a hundred years later they found out that drinking the purified water from the nanotechnology actually made some things worse? So it is a big debate on any scientific level." (Cristina, Eng/Env #1, Female, Asian, 26–35)

"We are stupid if we do not know that when something new comes out today [risk] research has to be a priority in the sense of you know as much as we can. Yeah we are not going to catch it all, we are not going to you know be able to say you know 10 years from

> now that something might happen, but in the general sense of it—is it safe publicly right now? (Nicole, Eng/Env #1, Female, White, 46–55)

US participants display in general a robust confidence in the American research and development enterprise to resolve uncertainty through research—they endorse a can-do belief in the solvability of these problems, even while acknowledging limits of knowledge in the current context. Risk perception research would argue that such trust in science and its management is much more important in the outcome of public response and risk perception than a particular risk finding in scientific research.[46] As shown in these examples, however, for a majority of participants there is concern about the ethics of trading off short-term benefits against potential long-term harms.

DISCUSSION AND CONCLUSION

Responsible development of nanotechnologies in the United States necessitates attention to the concerns, beliefs, ideas, and perceptions of diverse publics. In our deliberative workshops on nanotechnologies for energy and environment, we were able to convene diverse groups of citizens to discuss and facilitate their deliberation of the current and potential future ethical, social, and legal issues raised by new nanotechnologies for energy and environment. Normative ethical values were espoused at many junctures throughout the meetings, and cultural meanings were explored in small- and large-group sessions with particular attention to multiple perspectives rather than subsuming all under majority views.

In the 2009 CNS–UCSB deliberative workshops we encountered a number of conflicting views, some anticipated by literatures on risk perception and environmental decision making, and some reflecting the upstream context for these nanotechnological discussions and, perhaps, new aspects about nanotechnologies *or* the particular moment in time at which they are emerging (e.g., the economic crisis, massive job loss, military action, global distributed innovation system, lack of clear leadership on EH&S or other aspects of risk and safety). Strong perceived urgency seems among these participants to be driving toward attenuated risk views and to override otherwise strongly held views about governance and procedural justice, so it seems likely to temper deployment of ethical, value-driven decision making. Economic inequality and the current economic crisis themselves in these accounts also appear to significantly impede attention to transnational, global concerns, longer term benefits, and concern with collective benefits or the commons.

In this chapter, we have discussed some of the concerns and hopes raised by people around nanotechnology, to give a picture of a more bottom-up "nanoethics." In general, people were hopeful about the benefits of technology but wanted consistent and reliable research to minimize uncertainties around potentially negative or harmful effects, a sentiment expressed across gender and race. There was also a concern, particularly pronounced by nonwhite and female participants, that the benefits of nanotechnology would be distributed unequally. Many also expressed skepticism about the ability of companies and governing institutions to ensure public safety over profits, prompting calls for labeling. Thus while participants were hesitant about the ability of people to govern new technologies, they laid out a vision of the responsible development of nanotechnology, in which uncertain risks are minimized before marketing, benefits are more equitably available and distributed, and responsible institutions proactively balance profits with public health. Thus, for nanotechnology to be developed in this manner is a challenge, but doing so would build up the public's trust, and arguably make its members more invested in the possibilities and potentials of these new technologies. This to us would be democratic nanoethics.

CONTRIBUTORS

Richard Barrett, PhD, is emeritus professor of economics at the University of Montana. He is editor of *International Dimension of the Environmental Crisis* (Westview Press, 1982) and coauthor with T. Power of *Post-Cowboy Economics* (Island Press, 2002).

Albert Borgmann, PhD, is regents professor of philosophy at the University of Montana. His special area is the philosophy of society and culture with particular emphasis on technology. Among his publications are *Technology and the Character of Contemporary Life* (University of Chicago Press, 1984), *Crossing the Postmodern Divide* (University of Chicago Press, 1992), *Holding on to Reality: the Nature of Information at the Turn of the Millennium* (University of Chicago Press, 1999), and *Real American Ethics* (University of Chicago Press, 2006).

Donald A. Brown, Esq., is associate professor of environmental ethics, science, and law at Penn State and program director of the

Collaborative Program on the Ethical Dimensions of Climate Change. Mr. Brown has lectured and written widely about the need to integrate science, economics, and ethics in decision making. His most recent book is *American Heat: Ethical Problems with the United States' Response to Global Warming*. He also recently coedited *Sustainable Development: Science, Ethics, and Public Policy* with John Lemons.

David Castle, PhD, is Canada Research Chair in Science and Society at University of Ottawa. He is an associate professor in the Department of Philosophy and holds a cross-appointment to the faculty of law (Common Law Section). He is coauthor (with Cheryl Cline et. al.) of *Science, Society, and the Supermarket: The Opportunities and Challenges of Nutrigenomics* and editor (with Michael Ruse) of *Genetically Modified Foods: Debating Biotechnology*.

Julian Culp holds a BA in philosophy and economics from the University of Bayreuth (Germany) and an MA in political and economic philosophy from the University of Bern (Switzerland). For Alliance Sud, the Swiss Alliance of Development Organizations, he published with Bruno Stöckli *Understanding Participatory Poverty Reduction Strategy Monitoring Systems* (2007).

Nicole Hassoun, PhD, is an assistant professor in philosophy at Carnegie Mellon University. Her articles appear in journals such as the *American Philosophical Quarterly, Public Affairs Quarterly, Environmental Ethics, The American Journal of Bioethics, Journal of Moral Philosophy,* and *Utilitas*.

Barbara Herr Harthorn, PhD, is associate professor of feminist studies, anthropology, and sociology at University of California, Santa Barbra. Her work is published in a variety of social science, medical care, public health, environmental science and technology, technology and society, and nanoscience journals. She is editor (with Laury Oaks) of *Risk, Culture, and Health Inequality: Shifting Perceptions of Danger and Blame* (2003).

Tyronne Martin is a PhD student in chemistry at University of California, Santa Barbra. His research interests include protein design and engineering, along with drug-microtubule interactions and stability.

Bill McKibben is an American environmentalist and writer and a scholar in residence at Middlebury College. He is the author of *The End of Nature; The Age of Missing Information; Hope, Human and Wild; The Comforting Whirlwind: God, Job, and the Scale of Creation; Maybe One; Long Distance: A Year of Living Strenuously; Enough; Wandering Home*; and *Deep Economy: The Wealth of Communities and the Durable Future*. March 2008 saw the publication of *The Bill McKibben Reader*, a collection of forty-four essays written for various publications over the past twenty-five years.

Christopher J. Preston, PhD, is an associate professor of philosophy in the Department of Philosophy and a fellow at the Mansfield Ethics and Public Affairs Program at the University of Montana. He is author of *Grounding Knowledge: Environmental Philosophy, Epistemology, and Place* (University of Georgia Press, 2003) and *Saving Creation: Nature and Faith in the Life of Holmes Rolston, III* (Trinity University Press, 2009), as well as coeditor of *Nature, Values, and Duty: Life on Earth with Holmes Rolston, III* (Springer, 2007). He writes on topics in environmental philosophy, feminist ethics, and emerging technologies.

Wendy Parker is assistant professor of philosophy at Ohio University. Her research focuses primarily on questions in the philosophy of science, particularly questions about the nature and use of computer simulation models. She is also very interested in how science can contribute to, and sometimes obstruct, the development of public policy.

Jennifer Rogers, PhD, is assistant professor of sociology at Long Island University. Her dissertation "The Ma(i)ze of Globalization: Free Trade, Gender, and Resistance in Oaxaca" tells the multidimensional story of corn in Mexico. Jennifer's research centers the voices of indigenous women and their relationship to culture, the environment, and food production within the world economy.

Bernrd Rollin, PhD, is a University Distinguished Professor and is a professor of philosophy, biomedical sciences, and animal sciences at Colorado State University. In addition to numerous articles in the history of philosophy, philosophy of language, ethics, and bioethics, he is the author of *Natural and Conventional Meaning* (1976), *Animal Rights and Human Morality* (1981, 1993, and 2006), *The Unheeded Cry: Animal Consciousness, Animal Pain, and Scientific Change* (1988, 1998), *Farm Animal Welfare* (1995), *The Frankenstein Syndrome* (1995), and *Science and Ethics* (2006). He is the editor of *The Experimental Animal in Biomedical Research* (1989, 1995).

Ronald Sandler, PhD, is an associate professor of philosophy in the Department of Philosophy and Religion. He is the author of *Character and Environment* (Columbia University Press) and *Nanotechnology: The Social and Ethical Issues* (Woodrow Wilson Center), as well as coeditor of *Environmental Justice and Environmentalism* (MIT Press) and *Environmental Virtue Ethics* (Rowman and Littlefield).

Dane Scott II, PhD, is director of the Mansfield Ethics and Public Affairs Program and associate professor, Department of Society and Conservation, College of Forestry and Conservation at the University of Montana.

Christine Shearer, PhD, is a postdoctoral scholar at the Center for Nanotechnology in Society at University of California, Santa Barbara. She is managing editor of *Conducive,* and author of the forthcoming book, *Kivalina: A Climate Change Story* (Haymarket Books, 2011).

Paul B. Thompson, PhD, holds the W. K. Kellogg Chair in Agricultural, Food, and Community Ethics at Michigan State University. He has published a number of volumes and papers on the philosophical and cultural significance of farming, notably *The Spirit of the Soil: Agriculture and Environmental Ethics* (1995) and *The Agrarian Roots of Pragmatism* (2000).

Clark Wolf, PhD, is director of bioethics and associate professor of philosophy at Iowa State University. As director of the bioethics program at Iowa State University, he has organized symposia and conferences on a wide variety of subjects, including global climate change, the politicization of science, coexistence of organic and GM crops, intellectual property, and others.

NOTES

CHAPTER 1. DEBATING SCIENCE: ETHICS EDUCATION AND DELIBERATION

1. Douglas Walton's books and articles on pragmatics and critical reasoning provide the logical foundations for the pedagogy I am advocating. Walton's "new dialectics" marks a significant advance in the field of informal logic and critical reasoning. See D. Walton, *Practical Reasoning: Goal-Driven, Knowledge-Based, Action-Guiding Argumentation* (Savage, MD: Rowman & Littlefield, 1990) and D. Walton, *The New Dialectic: Conversational Contexts of Argument* (Toronto: University of Toronto Press, 1988).

2. Andrew E. Dessler and Edward A. Parson, *The Science and Politics of Global Climate Change* (Cambridge: Cambridge University Press, 2006), p. 18.

3. Michel de Montaigne, *An Apology for Raymond Sebond*, trans. M. A. Screech (New York: Penguin Classics, 1988).

4. Paul B. Thompson, "Bioethics Issues in a Biobased Economy," in *Genetically Modified Foods: Debating Biotechnology*, ed, Michael Ruse and David Castle (Amherst, NY: Prometheus Books. 68–76), p. 73.

5. Douglas Walton's work on practical reason provides some of the inspiration behind this section. In addition, this section draws on several basic insights found in John Dewey's work on logic and critical thinking.

6. In real life, deliberations are often what Douglas Walton calls "mixed discourse." By this, Walton means that other forms of dialogue are mixed in with deliberation and play a supporting role. It is often the case that at points in a deliberation there is a "dialogical shift" to another type of dialogue. More specifically, participants will sometimes engage in negotiation, inquiry, or even debate, but these forms of dialogue are subordinate to the overall deliberative process. Further, they should serve to support the overall goal of deliberation, to decide what to do. See D. Walton, *Practical Reasoning*, and D. Walton, *The New Dialectic*.

7. Daniel Hillel, *Out of the Earth: Civilization and the Life of the Soil* (Berkeley: University of California Press, 1992), p. 261.

8. David Tilman, "The Greening of the Green Revolution," *Nature* 396 (1998): 211–12.

9. It must be noted that philosophy of technology has pointed out in recent decades that new technologies cannot be reduced to mere tools, or means to an end. They can alter ways of life in profound ways. This insight would somehow need to be incorporated into deliberations.

10. Aristotle, *Politics*, trans. E. Barker (London: Oxford University Press, 1972).

11. Cass R. Sunstein, *Infotopia: How Many Minds Produce Knowledge* (Oxford, UK: Orxford University Press, 2006), p. 55.

12. Ibid., p. 51.

13. Ibid., p. 96.

14. Ibid., p. 81.

15. Ibid., p. 65.

16. Ibid., p. 67.

17. Ibid., p. 69.

18. Ibid., p. 67.

19. Ibid., p. 72.

20. Jay R. Wallace, "Practical Reason," *Stanford Encyclopedia of Philosophy*, November 6, 2008, http://www.plato.stanford.edu/entries/practical-reason/ (accessed July 10, 2009).

21. Martha C. Nussbaum, "Human Functioning and Social Justice: In Defense of Aristotelian Essentialism," *Political Theory* 20, no. 2 (1992): 219.

22. Andre Comte-Sponville, *A Small Treatise on the Great Virtue* (New York: Henry Holt, 1996), p. 32.

23. Robert Talisse, *Democracy after Liberalism: Pragmatism and Deliberative Politics* (New York: Routledge, 2005), pp. 115–22.

24. Ibid., p. 113.

CHAPTER 2. OVERCOMING SCIENTIFIC IDEOLOGY

1. K. Black, "Scientific Illiteracy in the US," Cedars-Sinai Neurosciences Report (2004).

2. C. Dean, "Science Savvy? In the US, Not Much," *New York Times*, August 30, 2005.

3. J Sacks, "The American Anti-Intellectual Threat," *Business World*, September 25, 2008.

4. W. V. Lumb, *Small Animal Anesthesia* (Philadelphia: Lea & Febiger, 1963).

5. W. V. Lumb and E. W. Jones, *Veterinary Anesthesia* (Philadelphia: Lea & Febiger, 1973).

6. R. L. Kitchell and H. H. Erikson, eds., *Animal Pain: Perception and Alleviation* (Bethesda, MD: American Physiological Society, 1983).

7. L. Davis, "Species Differences in Drug Disposition as Factors in Alleviation of Pain," in Kitchell and Erickson, eds., *Animal Pain*, pp. 161–78.

8. B. E. Rollin, *The Frankenstein Syndrome: Ethical and Social Issues in the Genetic Engineering of Animals* (New York: Cambridge University Press, 1995).

9. A. J. Ayer, *Language, Truth, and Logic* (London: Victor Gollanez, 1946).

10. L. Wittgenstein, "Lecture on Ethics," *Philosophical Review* 74 (1965): 3–12.

11. W. T. Keeton and J. L. Gould, *Biological Science* (New York: W. W. Norton, 1986).

12. Ibid.

13. S. Mader, *Biology: Evolution, Diversity, and the Environment* (Dubuque, IA: W. M. C. Brown, 1987).

14. R. M. Marks and E. S. Sacher, "Under-Treatment of Medical Inpatients with Narcotic Analgesics," *Annals of Internal Medicine* 79 (1973): 173–81.

15. B. E. Rollin, *Science and Ethics* (New York: Cambridge University Press, 2006).

16. G. Gaskell, "Europe Ambivalent on Biotechnology," *Nature* 387 (1997): 845–47.

CHAPTER 3. OVERCOMING PHILOSOPHOBIA: A FEW ETHICAL TOOLS FOR THE SCIENCE DEBATES

1. Kristin Shrader-Frechette, *The Ethics of Scientific Research* (Savage, MD: Rowman & Littlefield, 1994).

2. Occasionally, it is possible to arrive at a threshold beyond which that rule need no longer be obeyed. For example, there might be a general obligation to save endangered species from extinction. However, once a certain cost threshold is reached and the money being spent to save that species exceeds a certain amount, then the obligation to save the species can be waived.

3. Again, for some deontologists there are thresholds beyond which lying might become okay. Telling a lie to save the lives of a dozen innocents hiding in your basement may be acceptable, even for a deontologist.

4. A deranged consequentialist might seek to create the consequences that create the most pain. This would still be consequentialism even though the particular consequences being sought are (presumably) not morally praiseworthy.

5. One way to avoid the conclusion is to embrace a type of "rule consequentialism" in which behavior is based on general rules demonstrated over time to have desirable states of affairs as their consequences. Torturing innocents, if adopted as a rule, does not generally promote good consequences so the practice should be prohibited.

6. See note 1.

7. Susan Lang, "Cornell Ecologist's Study Finds That Producing Ethanol and Biodiesel From Corn and Other Crops Is Not Worth the Energy," *Cornell News Service*, July 5, 2005, http://www.news.cornell.edu/stories/july05/ethanol.toocostly.ssl.html (accessed June 24, 2010).

8. Aldo Leopold, *A Sand County Almanac* (Ballantine Books, 1970), p. 196.

9. J. Baird Callicott, "Moral Monism in Environmental Ethics Defended," *Journal of Philosophical Research* 19 (1994): 51–60 and "The Case Against Moral Pluralism," in *Beyond the Land Ethic: More Essays in Environ-*

mental Philosophy, 143–69 (Albany: State University of New York Press, 1999).

10. Christopher J. Stone, *Earth and Other Ethics: The Case for Moral Pluralism* (New York: Harper & Row, 1988); Andrew Light, "The Case Against Moral Pluralism," in *Environmental Ethics: An Anthology*, ed. Andrew Light and Holmes Rolston III (Oxford: Blackwell, 2003).

11. Bryan Norton, "Practical Philosophy vs. Applied Philosophy: Toward an Environmental Policy Integrated According to Scale," in *Environmental Philosophy and Environmental Activism*, eds. L. Embree and D. Marietta, 125–47 (Lanham, MD: Rowman and Littlefield, 1995).

12. Ibid., p. 130

13. Ibid., p. 135

CHAPTER 4. BRIDGING THE GAP: GLOBAL JUSTICE IN HEALTH RESEARCH

1. Nicole Hassoun would like to thank Stanford's Center for Ethics in Society and the United Nations' World Institute for Development Economics Research for support during the course of this project as well as the Falk Foundaiton. Julian Culp and Nicole Hassoun would like to thank Ayelet Banai and Miriam Ronzoni for their comments on previous drafts of this article.

2. We use the terms "developed" and "developing" countries throughout. This distinction is equivalent to the World Bank's classification of "low and middle income" and "high income" countries. See World Bank, "Country Classification," Data (2010), http://www.data.worldbank.org/about/country-classifications (accessed June 2010).

3. Scientific inquiry is meant to include research in the natural and social sciences. We also acknowledge that ethicists can do science and vice versa, though the fact that it is difficult to do either well may give individuals a reason to specialize and collaborate with specialists in other disciplines.

4. Commission on Health Research for Development, *Health Research—Essential Link to Equity in Development* (New York: Oxford University Press, 1990), http://www.hsph.harvard.edu/health-research/files/essentiallinktoequityindevelopment.pdf (accessed June 2010).

5. The GDB is measured in disability adjusted life years (DALYs).

6. The Commission on Health Research for Development, *Health Research*, p. 29. The data refers to the year 1986.

7. Adapted from ibid., p. 30.

8. The creation of the Global Forum for Health Research was recommended by the World Health Organization's Ad Hoc Committee on Health Research Relating to Future Intervention Options in its report *Investing in Health Research and Development* (Geneva: World Health Organization, 1996), http://www.who.int/tdr/svc/publications/tdr-research-publications/investing-in-health (accessed June 2010).

9. Global Forum for Health Research, *The 10/90 Report on Health Research 1999*, http://www.globalforumhealth.org/Media-Publications/Publications/10-90-Report-on-Health-Research-1999 (accessed June 2010). The numbers refer to estimates that in 1990 the top twenty diseases and risks of the world were accountable for 90 percent of the GDB. However, only 10 percent of global health research was directed to them.

10. Thomas Pogge, "Could Globalization Be Good for World Health?" *Global Justice: Theory Practice Rhetoric* 1, no. 1 (2008): 7. The study of the Tufts Center for the Study of Drug Development, "Drug Approvals for Neglected Diseases Increase along with More R&D Funding," *Tufts CSDD Impact Report* (November 6, 2009), found that between 1999 and 2008 the R&D funding on neglected diseases has increased twenty-five fold.

11. Many health problems in developing countries could be ameliorated with existing medications. Nevertheless, few medications exist to treat some of the world's worst health problems. Reducing the health-research gap may yield new life-saving treatments or less expensive versions of existing medications.

12. The table is adapted from Global Forum for Health Research, *Monitoring Financial Flows for Health Research 2006—The Changing Landscape of Health Research for Development* (Geneva: Global Forum for Health Research, 2006), p. 90, http://www.globalforumhealth.org/Media-Publications/Publications/(view)/series/(series)/4 (accessed June 2010).

13. Pierre Chirac and Els Torreele, "Global Framework on Essential Health R&D," *Lancet* 367 (May 13, 2006): 1560–61, p. 1560. Earlier findings report that of the 1,223 newly commercialized chemical entities between 1975 and 1997 only 13 were for tropical diseases, i.e., diseases that are predominantly found in developing countries. See Bernard Pecoul, Pierre Chirac, Patrice Trouiller, and Jacques Pinel, "Access to Essential Drugs in Poor Countries: A Lost Battle?" *Journal of the American Medical Association* 281, no.4 (1999): 361–67.

14. The numbers refer to the year 2007. See United Nations Educational, Scientific, and Cultural Organization, Institute for Statistics Data

Center, "Regional Totals for R&D Expenditure (GERD) and Researchers, 2002 and 2007," *Public Report on Science and Technology*, http://www.stats.uis.unesco.org/unesco/ReportFolders/ReportFolders.aspx?IF_ActivePath=P,54&IF_Language=eng (accessed June 2010).

15. Ibid.

16. In the member countries of the Organisation for Economic Co-operation and Development (OECD), 69.8 percent of total R&D was executed by the industry in 2008 (or latest available year). See OECD, *Main Science and Technology Indicators: Volume 2010/1*, p. 18, http://www.oecd.org/dataoecd/9/44/41850733.pdf (accessed June 2010).

17. Ibid.

18. In 2008 the US health R&D investments totaled 29.265 billion (29,265 million) US dollars; see National Science Board, chapter 4, "Research and Development: National Trends and International Linkages," *Science and Engineering Indicators 2010* (Arlington, VA: National Science Foundation, 2010), p. 22, http://www.nsf.gov/statistics/seind10/ (accessed June 2010). Over a billion (1,258 million) thereof were spent on R&D for neglected diseases; see Mary Moran et al., *Research and Development: New Times, New Trends* (Sydney: George Institute, 2009), p. 55, http//www.georgeinstitute.org/sites/default/files/pdfs/G-FINDER_2009_Report.pdf (accessed June 2010).

19. Mary Moran et al., *Neglected Disease Research and Development*, pp. 55, 67. The United States spent roughly ten times more than the second largest public spender, the European Commission, on R&D for neglected diseases in 2008; see ibid., p. 55. The US National Institutes of Health, which spends 85.7 percent of the total health R&D of the US government, is the organization that contributed more to R&D for neglected diseases than any other public or private (for-profit or not-for-profit) body in 2008; see ibid., p. 67.

20. World Bank, *World Development Report 2008—Agriculture for Development* (Washington, DC: World Bank, 2007), http://www.siteresources.worldbank.org/INTWDR2008/Resources/WDR_00_book.pdf (accessed June 2010). See especially chapter 7.

21. Ibid.

22. Ibid.

23. Ibid.

24. To consider whether this is so, some moral theories ask, for instance, how the organization of scientific research affects fundamental human interests. For ethicists that argue that the health-research gap is morally problematic, see James Flory and Philip Kitcher, "Global Health

and the Scientific Research Agenda," *Philosophy and Public Affairs* 32, no. 1 (2004): 36–65, and Thomas Pogge, "Human Rights and Global Health: A Research Program," *Metaphilosophy* 36, nos. 1 and 2 (2005): 182–209.

25. Aidan Hollis and Thomas Pogge, *The Health Impact Fund: Making New Medicines Accessible for All* (Incentives for Global Health, 2008): http://www.yale.edu/macmillan/igh/hif_book.pdf (accessed June 2010). Hollis and Pogge's proposal is similar in some ways to the proposal advanced in Tim Hubbard and James Love, "A New Trade Framework for Global Healthcare R&D," *PLoS Biology* 2, no. 2 (2004), http://www.plosbiology.org/article/info:doi/10.1371/journal.pbio.0020052 (accessed June 2010).

26. Pogge, "Human Rights and Global Health," p. 191. The program might cost quite a bit more to really make an impact, as drug companies report average R&D costs in the hundreds of millions; see Marcia Angell, *The Truth about the Drug Companies: How They Deceive Us and What to Do about It* (New York: Random House, 2004), p. 43.

27. Hollis and Pogge, *The Health Impact Fund*, p. 10.

28. Note that in 1970 the United Nations General Assembly adopted resolution 2626, which demanded that developed countries increase their official development assistance (ODA) to 0.7 percent of their GNI. For 2010 the OECD estimates that the ODA/GNI ratio will be 0.32 percent on average. See "Development Aid Rose in 2009 and Most Donors Will Meet 2010 Aid Targets," http://www.oecd.org/document/11/0,3343,en_2157 1361_44315115_44981579_1_1_1_1,00.html (accessed June 2010).

29. Carol Mimura, "Technology Licensing for the Benefit of the Developing World: UC Berkeley's Socially Responsible Licensing Program," *Journal of the Association of University Technology Managers* 27, no. 2, pp. 15–28, http://www.autm.net/AM/Template.cfm?Section=Global_Health&Template=/CM/ContentDisplay.cfm&ContentID=3718 (accessed June 2010).

30. Ibid.

31. Ibid., p. 19.

32. John Lauerman, "Harvard among Six Schools Urging Drug Access for Poor," Bloomberg (November 9, 2009), http://www.bloomberg.com/apps/news?pid=newsarchive&sid=aa23AHBWnxew (accessed June 2010).

33. World Health Assembly, "Global Strategy and Plan of Action on Public Health, Innovation, and Intellectual Property," WHA resolution 61.21 (2008), http://www.apps.who.int/gb/ebwha/pdf_files/A61/A61_R21-en.pdf (accessed June 2010).

34. This issue, "Transfer of technology," is Element 4 of the global strategy; see World Health Assembly 61.21, p. 14.

35. See the report of the World Health Organization Commission on Intellectual Property Rights, Innovation, and Public Health, *Public Health—Innovation and Intellectual Property Rights* (Geneva: World Health Organization, 2006), p. 174, http://www.esocialsciences.com/data/articles/Document11442006230.6463587.pdf (accessed June 2010).

36. Ibid., p. 172.

37. Cheri Grace, *The Effect of Changing Intellectual Property on Pharmaceutical Industry Prospects in India and China: Considerations for Access to Medicine* (London: Department for International Development, 2004), p. 18, http://www.dfid.gov.uk/pubs/files/indiachinadomproduce.pdf (accessed June 2010).

38. Warren Kaplan and Richard Laing, "Local Production of Pharmaceuticals: Industrial Policy and Access to Medicines," *Health Nutrition and Population Discussion Papers* (Washington, DC: World Bank, 2005), p. iii, http://www.siteresources.worldbank.org/HEALTHNUTRITIONANDPOPULATION/Resources/281627-1095698140167/KaplanLocalProductionFinal.pdf (accessed June 2010).

39. For GSK's technology transfer initiatives, see GSK, "'Technology Transfer,' Capacity Building, and the Developing World," *Global Public Policy Issues* (September 2007), http://www.gsk.com/policies/GSK-on-technology-transfer-capacity-building.pdf (accessed June 2010).

40. Médecins Sans Frontières and Drugs for Neglected Diseases Working Group, *Fatal Imbalance—The Crisis in Research and Development for Neglected Diseases* (Geneva: Médecins Sans Frontières, 2001), http://www.msf.org/source/access/2001/fatal/fatal.pdf (accessed June 2010).

41. Ibid. pp. 2, 4.

42. Helen Frankish, "Initiative Launched to Develop Drugs for Neglected Diseases," *Lancet* 362 (July 12, 2003): 135.

CHAPTER 5. INTELLECTUAL LIBERTY AND THE PUBLIC REGULATION OF SCIENTIFIC RESEARCH

1. A. Revkin, "Bush Aide Softened Greenhouse Gas Links to Global Warming," *New York Times*, June 8, 2005, http://www.nytimes.com/2005/06/08/politics/08climate.html?scp=1&sq=Philip%20Cooney&st=cse (accessed August 1, 2010).

2. M. Nussbaum, *Liberty of Conscience* (New York: Basic Books, 2008).

3. M. Callon, "Is Science a Public Good?" *Science, Technology, and Human Values* 19, no. 4 (1994): 395–424; K. Stengel, J. Taylor, Claire Waterton, and B. Wynne, "Plant Sciences and the Public Good," *Science, Technology, and Human Values* 34, no. 3 (2009): 289–312.

4. C. Wolf, "Liberalism and Fundamental Constitutional Rights," *Arizona Law Review* 37, no. 1 (1995): 185–95; J. Feinberg, *Problems at the Roots of Law* (New York: Oxford University Press, 2003).

5. Wolf, "Liberalism and Fundamental Constitutional Rights."

6. J. S. Mill, *On Liberty* (Cambridge: Hackett Publishing Co., 1859/1982).

7. J. Feinberg, *Harm to Others* (New York: Oxford University Press, 1984); Wolf, "Liberalism and Fundamental Constitutional Rights."

8. T. Ferris, *The Science of Liberty* (New York: Harper Collins, 2010) p. 13.

9. J. Milton, "Paradise Lost," in *The Complete Poetry of John Milton*, ed. John T. Shawcross (New York: Doubleday, 1674/1963) p. 391.

10. Milton, *Paradise Lost*, bk. 7, 120–25

11. Ibid., lines 126–30

12. W. Berry, "Faustian Economics," *Harpers Magazine*, May 8, 2008, http://www.harpers.org/archive/2008/05/0082022 (accessed August 2010).

13. Ibid.; McKibben, *Enough* (New York: Henry Holt & Co., 2003).

14. Letter to the editor, *New York Times*, May 26, 2005.

15. See J. Rawls, "The Idea of Public Reason Revisited," in *Collected Papers* (Cambridge: Harvard University Press, 1997/1999), pp. 573–615.

16. J. Rawls, *Political Liberalism* (New York: Columbia University Press, 1993), p. 137.

17. Rawls's definition does leave open the possibility that we might not be able to justify reasonable requirements to people who are unreasonable or irrational. This leaves open the possibility that those who don't agree might simply be classed as "unreasonable," but Rawls takes steps to prevent this subversive interpretation of his view. See Rawls, *Political Liberalism* and C. Wolf, "Fundamental Rights, Reasonable Pluralism, and the Moral Commitments of Political Liberalism," in *The Idea of a Political Liberalism*, ed. V. Davion and C. Wolf (New York: Rowman & Littlefield, 2000).

18. Revkin, "Bush Aide Softened Greenhouse Gas Links."

19. Nussbaum, *Liberty of Conscience.*

20. J. Feinberg, "Not with My Tax Money: The Problem of Justifying Government Subsidies for the Arts," in *Problems at the Roots of the Law:*

Essays in Political and Legal Theory (New York: Oxford University Press, 2003), pp.103–124.

21. J. Rawls, *Political Liberalism*, p. 137.

CHAPTER 7. COMMUNICATING SCIENCE: MORAL RESPONSIBILITY IN THEORY AND PRACTICE

1. See Peter Singer, "Famine, Affluence, and Morality," *Philosophy and Public Affairs* 1 (1972): 229–43.

2. See e.g., Heather E. Douglas, "The Moral Responsibilities of Scientists: Tensions between Autonomy and Responsibility," *American Philosophical Quarterly* 40 (2003): 59–68.

3. This does not mean that the research findings should NOT be communicated; P1 is silent on what should happen when (b) is not met.

4. S. Solomon et al., *Climate Change 2007: The Physical Science Basis: Contribution of Working Group I to the Fourth Assessment Report of the Intergovernmental Panel on Climate Change* (New York: Cambridge University Press, 2007).

5. Some increase in the average near-surface temperature of Earth's atmosphere is fully expected. Just how large that increase will be by the end of the twenty-first century under different emission scenarios and on smaller spatial scales is less clear.

6. Heather E. Douglas, *Science, Policy, and the Value-Free Ideal* (Pittsburgh: University of Pittsburgh Press, 2009); see especially chapter 4. See also Richard Rudner, "The Scientist qua Scientist Makes Value Judgments," *Philosophy of Science* 20 (1953): 1–6.

7. Even if we reject that scientists must decide whether to accept or reject hypotheses and argue instead that scientists only need to assign probabilities to such hypotheses (e.g., Richard Jeffrey, "Valuation and Acceptance of Scientific Hypotheses," *Philosophy of Science* 23 [1956]: 237– 46), there are still questions about the conditions under which it is appropriate to communicate such probability assignments to decision makers. For related discussion, see Wendy S. Parker, "Whose Probabilities? Predicting Climate Change with Ensembles Models," *Philosophy of Science* 77 (2010): 985–97, and Douglas, *Science, Policy, and the Value-Free Ideal*, chapter 3 and p. 85.

8. See e.g., the discussion of "scientific accuracy" vs. "communicative accuracy" in John Beatty, "Masking Disagreements among Experts," *Episteme* 3 (2006): 52–67.

9. See also American Psychological Association (APA), "Psychology and Global Climate Change: Addressing a Multi-Faceted Phenomenon and Set of Challenges," *Report of the American Psychological Association Task Force on the Interface between Psychology and Global Climate Change*, 2010, http://www.apa.org/science/about/publications/climate-change.aspx (accessed July 3, 2010).

10. On a related note, science reporter Andrew Revkin has suggested that ordinary scientific discourse related to climate change (e.g., debates over details) can leave the public with the mistaken impression that very little is known about the issue (see Andrew C. Revkin, "Climate Experts Tussle over Details. Public Gets Whiplash," *New York Times*, July 29, 2008).

11. For a climate scientist's perspective on politicization and scientific communication, see Stephen H. Schneider, *Science as a Contact Sport: Inside the Battle to Save Earth's Climate* (Washington, DC: National Geographic Books, 2009), especially chapter 7.

CHAPTER 8. IS SUSTAINABILITY WORTH DEBATING?

1. Rodger W. Bybee, "Planet Earth in Crisis: How Should Science Educators Respond?" *American Biology Teacher* 53, no. 3 (1991).

2. International Union for the Conservation of Nature and Natural Resources (IUCN), *World Conservation Strategy* (Gland, Switzerland: IUCN-UNDP-WWF, 1980).

3. World Commission on Environment and Development, *Our Common Future* (New York: Oxford University Press, 1987).

4. Carl Mitcham, "The Concept of Sustainable Development: Its Origins and Ambivalence," *Technology in Society* 17, no. 3 (1995).

5. Paul B. Thompson, *The Agrarian Vision: Sustainability and Environmental Ethics* (Lexington: University Press of Kentucky, 2010).

6. Paul B. Thompson, *The Spirit of the Soil: Agriculture and Environmental Ethics* (London and New York: Routledge, 1995).

7. D. Jamieson, "Sustainability and Beyond," *Ecological Economics* 24 (1998).

8. Karl-Hennk Robèrt et al., "A Compass for Sustainable Development," *International Journal of Sustainable Development & World Ecology* 4,

no. 2 (1997); Robert Costanza et al., "Managing Our Environmental Portfolio," *BioScience* 50, no. 2 (2000); Maureen Rogers and Roberta Ryan, "The Triple Bottom Line for Sustainable Community Development," *Local Environment: The International Journal of Justice and Sustainability* 6, no. 3 (2001).

9. W. C. Clark, "Sustainability Science: A Room of Its Own," *Proceedings of the National Academy of Sciences of the United States of America* 104, no. 6 (2007); Robert Kates et al., "Sustainability Science" (December 2000) KSG Working Paper No. 00-018. Available at SSRN, http://.ssrn.com/abstract=257359 or doi:10.2139/ssrn.257359

10. Aiden Davison, *Technology and the Contested Meanings of Sustainability* (Albany: State University of New York Press, 2001); Julianne Lutz-Newton and Eric T. Freyfogle, "Sustainability: A Dissent," *Conservation Biology* 19 (2005).

11. Tom Athanasiou, "The Age of Greenwashing," *Capitalism, Nature, Socialism* 7, no. 1 (1996).

12. R. M. Solow, "Sustainability: An Economist's Perspective," in *Economics of the Environment: Selected Readings*, ed. R. Dorfman and N. Dorfman (New York: Norton, 1993); David Pearce, "Economics, Equity, and Sustainable Development," *Futures* 20, no. 6 (1988).

13. B. G. Norton, *Sustainability: A Philosophy of Adaptive Ecosystem Management* (Chicago: University of Chicago Press, 2005).

14. Elinor Ostrom, *Governing the Commons: The Evolution of Institutions for Collective Action* (Cambridge, UK: Cambridge University Press, 1990); Peter J. Taylor, *Unruly Complexity: Ecology, Interpretation, and Engagement* (Chicago: University of Chicago Press, 2005).

15. Beate Littig and Erich Griessler, "Social Sustainability: A Catch-Word between Political Pragmatism and Social Theory," *International Journal of Sustainable Development* 8 (2005); Aimee Shreck, Christy Getz, and Gail Feenstra, "Social Sustainability, Farm Labor, and Organic Agriculture: Findings from an Exploratory Analysis," *Agriculture and Human Values* 23, no. 4 (2006).

16. Amartya Sen, *The Standard of Living* (Cambridge: Cambridge University Press, 1987); Amartya Sen, "Editorial: Human Capital and Human Capability," *World Development* 25, no. 12 (1997).

17. I. Kawachi et al., "Social Capital, Income Inequality, and Mortality," *American Journal of Public Health* 87, no. 9 (1997); Robert D. Putnam, "Bowling Alone: America's Declining Social Capital," *Journal of Democracy* 6, no. 1 (1995): 65–78.

18. Markku Lehtonen, "The Environmental-Social Interface of Sus-

tainable Development: Capabilities, Social Capital, Institutions," *Ecological Economics* 49, no. 2 (2004).

19. Thompson, *The Spirit of the Soil;* Peter Marcuse, "Sustainability Is Not Enough," *Environment and Urbanization* 10, no. 2 (1998).

20. P. Allen and C. Sachs, "The Poverty of Sustainability: An Analysis of Current Positions," *Agriculture and Human Values* 9, no. 4 (1992).

21. Theodor W. Adorno, *The Culture Industry: Selected Essays on Mass Culture* (New York: Routledge, 1991); Max Horkheimer and Theodor W. Adorno, *Dialectic of Enlightenment: Philosophical Fragments,* ed. Gunzelin Schmid Noerr, trans. Edmund Jephcott (Stanford, CA: Stanford University Press, 2002); Herbert Marcuse, *One-Dimensional Man* (Boston: Beacon Press, 1964).

22. Elizabeth Barham, "Social Movements for Sustainable Agriculture in France: A Polanyian Perspective," *Society & Natural Resources: An International Journal* 10, no. 3 (1997); Laura Raynolds, Douglas Murray, and Andrew Heller, "Regulating Sustainability in the Coffee Sector: A Comparative Analysis of Third-Party Environmental and Social Certification Initiatives," *Agriculture and Human Values* 24, no. 2 (2007).

23. Julie Guthman, "The Polanyian Way? Voluntary Food Labels as Neoliberal Governance," *Antipode* 39, no. 3 (2007).

24. Ulrich Beck, Anthony Giddens, and Scott Lash, *Reflexive Modernization: Politics, Tradition, and Aesthetics in the Modern Social Order* (Stanford, CA: Stanford University Press, 1994); Peter Gundelach and Lars Torppe, "Social Reflexivity, Democracy, and New Types of Citizen Involvement in Denmark," in *Private Groups and Public Life: Social Participation, Voluntary Associations,* ed. Jan W. Van Deth (New York: Routledge, 1997).

CHAPTER 9. BIOTECHNOLOGY AND THE PURSUIT OF FOOD SECURITY

1. Food and Agriculture Organization of the United Nations (FAO), "Special Programme on Food Security," http://www.fao.org/spfs/about-spfs/frequently-asked-questions-spfs/en/ (accessed July 24, 2010)

2. Nature Editor, "How to Feed a Hungry World," *Nature* 466 (2010).

3. P. Pinstrup-Andersen, "Food Security: Definition and Measurement," *Food Security* 1 (2009).

4. Ibid.

5. Food and Agriculture Organization of the United Nations (FAO), *The State of Food Insecurity in the World* (Rome: FAO, 2006).

6. Center on Hunger and Poverty, *The Consequences of Hunger and Food Insecurity for Children* (Waltham, MA: Heller School for Social Policy and Management, Brandeis University, 2002).

7. Douglas DeWitt Southgate, Douglas H. Graham, and Luther G. Tweeten, *The World Food Economy* (Malden, MA: Blackwell, 2007).

8. Food and Agriculture Organization of the United Nations (FAO), "Rome Declaration on World Food Security," http://www.fao.org/docrep/003/w3613e/w3613e00.HTM (Accessed July 24, 2010).

9. Pinstrup-Andersen, "Food Security."

10. United Nations, "55/2. United Nations Millennium Declaration," http://www.un.org/millennium/declaration/ares552e.htm (accessed July 24, 2010).

11. World Bank, *World Development Report 2010: Development and Climate Change* (Washington, DC: World Bank, 2010).

12. United Nations Habitat, "Fact Sheet: Habitat for Global Development, Global Report on Human Settlements" (2009).

13. L. R. Brown and Earth Policy Institute, *Outgrowing the Earth: The Food Security Challenge in the Age of Falling Water Tables and Rising Temperatures*, 1st ed. (New York: W. W. Norton & Co., 2004).

14. United States Census Bureau, "International Data Base: Total Midyear Population for the World: 1950–2050," http://www.census.gov/ipc/www/idb/worldpop.php (accessed July 24, 2010).

15. J. Bryant, "Theories of Fertility Decline and the Evidence from Development Indicators," *Population and Development Review* 33 (2007).

16. M. Myrskyla, H-P. Kohler, and F. C. Billari, "Advances in Development Reverse Fertility Declines," *Nature* 460 (2009).

17. D. Grigg, "The Pattern of World Protein Consumption," *Geoforum* 26 (1995).

18. Food and Agriculture Organization of the United Nations (FAO), "World Calories," http://www.fao.org/economic/ess/chartroom-and-factoids/chartroom/93-world-calories-total/en/ (accessed July 24, 2010).

19. Dambisa Moyo, *Dead Aid* (London: Granta Books, 2009).

20. L. Smith, A. E. Obeid, and H. Jensen, "The Geography and Causes of Food Insecurity in Developing Countries," *Agricultural Economics* 22 (2000).

21. P. A. Sanchez, "A Smarter Way to Combat Hunger," *Nature* 458 (2009).

22. G. L. Denning et al., "Input Subsidies to Improve Smallholder Maize Productivity in Malawi: Toward an African Green Revolution," *PLoS Biology* 7 (2009).

23. P. A. Sanchez, G. L. Denning, and G. Nziguheba, "The African Green Revolution Moves Forward," *Food Security* 1 (2009).

24. J. Sachs, *The End of Poverty* (New York: Penguin, 2006).

25. J. A. Thompson, "The Role of Biotechnology for Agricultural Sustainability in Africa," *Philosophical Transactions of the Royal Society B* 363 (2008).

26. Ibid.

27. Ibid.

28. Ibid.

29. R. Paarlberg, *Starved for Science: How Biotechnology Is Being Kept out of Africa* (Cambridge: Harvard University Press, 2008).

30. V. Shiva and G. Bedi, *Sustainable Agriculture and Food Security: The Impact of Globalization* (New Delhi, Thousand Oaks: Sage Publications, 2002).

31. J. Sachs et al., "Monitoring the World's Agriculture," *Nature* 466 (2010).

32. K. Amman, "Integrated Farming: Why Organic Farmers Should Use Transgenic Crops," *New Biotechnology* 25, no. 2 (2008); K. Amman, "Why Farming with High-Tech Methods Should Integrate Elements of Organic Agriculture," *New Biotechnology* 25, no. 6 (2009).

33. Amman, "Why Farming with High Tech Methods Should Integrate Elements of Organic Agriculture."

34. Amman, "Integrated Farming."

35. J. Gressel, *Genetic Glass Ceilings: Transgenics for Crop Biodiversity* (Baltimore: Johns Hopkins University Press, 2008).

36. C. Juma and L. Yee-Cheong, *Innovation: Applying Knowledge in Development, UN Millennium Project Task Force on Science, Technology, and Innovation* (London: Earthscan, 2005).

37. World Bank, *World Development Report 2010*.

38. D. Castle et al., "Knowledge Management and the Contextualization of Intellectual Property Rights in Innovation Systems," *SCRIPT-ed: A Journal of Law, Technology, and Society* 7 (2010).

39. N. Gilbert, "Inside the Hothouses of Industry," *Nature* 466 (2010).

40. E. R. Gold et al., "The Unexamined Assumptions of Intellectual Property: Adopting an Evaluative Approach to Patenting Biotechnological Innovation," *Public Affairs Quarterly* 18 (2004); D. Castle, ed., *The Role of Intellectual Property Rights in Biotechnology Innovation* (Cheltenham, UK: Edward Elgar, 2009).

CHAPTER 11. SCIENCE, ETHICS, AND TECHNOLOGY AND THE CHALLENGE OF GLOBAL WARMING

1. Francis Bacon, *The Two Books of Francis Bacon: Of the Proficience and Advancement of Learning, Divine and Human* (Charleston, SC: BiblioLife, 2009 [1623]), p. 35.

2. Albert Borgmann, "Traditional Culture and Global Commodification," *Bulletin of the Boston Theological Institute* 8 (2008): 18–23.

3. Christopher D. Stone, *Should Trees Have Standing?* (Los Altos, CA: William Kaufmann, 1974 [1972]).

4. McKinsey and Company, "Reducing US Greenhouse Gas Emissions: How Much at What Cost?" December 2007, http://www.mckinsey.com/en/Client_Service/Sustainability/Latest_Thinking/Reducing_US_greenhouse_gas_emissions.aspx (accessed October 5, 2011).

5. Martin Gilens, "Inequality and Democratic Responsiveness," *Public Opinion Quarterly* 69, no. 5 (2005): 778–896.

6. Michael I. Norton and Dan Ariely, "Building a Better America—One Wealth Quintile at a Time," *Perspectives on Psychological Science* 6, no. 9 (2011): 9–12

7. Martin Heidegger, Die Technik und die Kehre (Pfullingen: Neske, 1962).

CHAPTER 12. TEN ETHICAL QUESTIONS THAT SHOULD BE ASKED OF THOSE WHO OPPOSE CLIMATE CHANGE POLICIES ON SCIENTIFIC GROUNDS

1. Intergovernmental Panel on Climate Change, "History," 2010, http://www.ipcc.ch/organization/organization_history.htm (accessed June 23, 2010).

2. Ibid.

3. Ibid.

4. Ibid.

5. Intergovernmental Panel on Climate Change, "Working Groups/Task Forces," 2010, http://www.ipcc.ch/working_groups/working_groups.htm (accessed June 23, 2010).

6. Intergovernmental Panel on Climate Change, 2010, http://www.ipcc.ch/publications_and_data/publications_and_data_reports.htm#1 (accessed June 23, 2010).

7. Intergovernmental Panel on Climate Change, "IPCC Fourth Assessment Report: Climate Change 2007, Working Group I, Summary for Policy Makers," 2007, http://www.ipcc.ch/publications_and_data/ar4/wg1/en/spm.html (accessed June 23, 2010).

8. For a good summary of the arguments most frequently made by climate change skeptics, see http://www.skepticalscience.com.

9. For examples of responses by mainstream scientists to climate skeptics' arguments, see http://www.realclimate.org and http://www.skepticalscience.com.

10. William L. Anderegga, James W. Prall, Jacob Harold, and Stephen H. Schneider, "Expert Credibility in Climate Change," *Proceedings of the National Academy of Sciences of the United States*, April 9, 2010, http://www.pnas.org/content/early/2010/06/04/1003187107.full.pdf+html (accessed June 23, 2010).

11. Peter T. Doran and Maggie Kendall Zimmerman, "Examining the Scientific Consensus on Climate Change," *EOS* 90, no. 3 (2009): 22–23.

12. Ibid.

13. Skeptical Science, "Getting Skeptical about Skeptics Science," 2010, http://www.skepticalscience.com/ (accessed June 23, 2010).

14. Ibid.

15. Stepan Rahmstorf et al., "Recent Climate Observations Compared to Projections," *Science* 316 (May 2007).

16. Lyndon Johnson, "Special Message to Congress on Conservation and Restoration of Natural Beauty," *American Presidency Project*, 1965, http://www.presidency.ucsb.edu/ws/index.php?pid=27285 (accessed June 23, 2010).

17. Robert White, "Oceans and Climate—Introduction," *Oceanus* 21, nos. 2–3 (1978).

18. National Research Council (NRC), *Carbon Dioxide and Climate: A Scientific Assessment* (Washington DC: National Academy Press, 1979).

19. Ibid.

20. Shardul Agrarwala and Stiener Anderson, "Indispensability and Indefensibility?: The United States in Climate Treaty Negotiations," *Global Governance* 5 (1999): 457–82.

21. Naomi Oreskes and Erik Conway, *Merchants of Doubt* (New York: Bloomsbury Press, 2010).

22. Ibid., pp. 186–90

23. Ibid., p. 190

24. For a description of corporate-funded campaigns that were created to convince citizens that the IPCC conclusions were not grounded in science, see ibid., and James Hogan, *Climate Cover-Up: The Crusade to Deny Global Warming* (Vancouver: Greystone Books, 2010).

25. Hogan, *Climate Cover-Up*, p. 32

26. Ibid.

27. Ibid., p. 12.

28. Ibid.

29. Ibid., p. 13.

30. Andrew C. Revkin, "Industry Ignored Its Scientists on Climate," *New York Times*, April 23, 2009.

31. Hogan, *Climate Cover-Up*, p. 65.

32. Ibid., p. 66.

33. Ibid., p. 81.

34. Ibid.

35. Andrew Rich, "US Think Tanks and the Intersections of Ideology, Advocacy, and Influence," *NIRA Review*, Winter 2001, pp. 54–59.

36. Ibid.

37. Michael Dolny, "What's in a Label?: Right-Wing Think Tanks Are Often Quoted, Rarely Labeled," *Extra!*, May/June 1998, http://www.fair.org/index.php?page=1425 (accessed June 23, 2010).

38. Ibid.

39. Ibid.

40. Chris Mooney, "Some Like It Hot: As the World Burns," *Mother Jones*, May/June 2005, http://motherjones.com/environment/2005/05/some-it-hot (accessed April 3, 2005).

41. Ibid.

42. Hogan, *Climate Cover-Up*, p. 73.

43. Ibid., p. 74.

44. S. Covington, "How Conservative Philanthropies and Think Tanks Transform US Policy," *Covert Action Quarterly*, Winter 1998, http://mediafilter.org/CAQ/caq63/caq63thinktank.html (accessed June 23, 2010).

45. David A. Fahrenthold and Juliet Eilperin, "In E-Mails, Science of Warming is Hot Debate," *Washington Post*, December 5, 2009.

46. Ibid.

47. Fred Pearce, "How the 'Climategate' Scandal Is Bogus and Based on Climate Sceptics' Lies," *Gaurdian*, February 9, 2010, http://www

.guardian.co.uk/environment/2010/feb/09/climategate-bogus-sceptics-lies (accessed June 23, 2010).

48. Sharon Begley, "Newspapers Retract Climategate Claims, but Damage Still Done," *Newsweek*, June 26, 2010, http://www.newsweek.com/blogs/the-gaggle/2010/06/25/newspapers-retract-climategate-claims-but-damage-still-done.html (accessed June 23, 2010).

49. Ibid.

50. Stephen Schneider, "Hammering Out a Deal for Our Future," *Huffington Post*, December 9, 2009, http://www.huffingtonpost.com/stephen-h-schneider/hammering-out-a-deal-for_b_386117.html (accessed June 23, 2010).

51. See e.g., Union of Concerned Scientists, "Human Finger Prints," 2010, http://www.ucsusa.org/global_warming/science_and_impacts/science/global-warming-human.html (accessed June 23, 2010).

52. Intergovernmental Panel on Climate Change (IPCC), "Statement by Working Group I of the Intergovernmental Panel on Climate Change on Stolen E-mails from the Climate Research Unit at the University of East Anglia, United Kingdom," December 2009, http://www.ipcc.ch/pdf/presentations/WGIstatement0412209.pdf (accessed June 23, 2010).

53. American Meteorological Society (AMS), "Impact of CRU Hacking on the AMS Statement on Climate Change," 2010, http://www.webcitation.org/5lnFDGhdZ (accessed June 23, 2010).

54. American Geophysical Union (AGU), "AGU Statement Regarding the Recent Release of E-mails Hacked from the Climate Research Unit at University of East Anglia," 2010, http://www.agu.org/news/archives/2009-12-08_hacked-emails-climate-researchshtml.shtml (accessed June 23, 2010).

55. American Association for the Advancement of Science (AAAS), "AAAS Reaffirms Statements on Climate Change and Integrity," 2010, http://www.aaas.org/news/releases/2009/1204climate_statement.shtml (accessed June 23, 2010).

56. House of Commons Science and Technology Committee (STC), "The Disclosure of Climate Data from the Climatic Research Unit at the University of East Anglia," 2010, http://www.climateprogress.org/wp-content/uploads/2010/03/HC387-IUEAFinalEmbargoedv21.pdf (accessed June 23, 2010).

57. Muir Russel, Geoffry Bullton, Peter Clarke, David Eyton, and James Norton, "Independent Climate Change E-mails Review, 2010, http://www.cce-review.org/pdf/FINAL%20REPORT.pdf (accessed June 23, 2010).

58. Henry C. Foley, Alan W. Scaroni, and Candice A. Yekel, "RA-10 Inquiry Report: Concerning the Allegations of Research Misconduct against Dr. Michael E. Mann, Department of Meteorology, College of Earth and Mineral Sciences, the Pennsylvania State University, February 3, 2010, http://www.research.psu.edu/orp/Findings_Mann_Inquiry.pdf (accessed June 23, 2010).

59. Naomi Oreskes, "Beyond the Ivory Tower: The Scientific Consensus on Climate Change," *Science* 306, no. 5702 (December 2004): 1686.

CHAPTER 13. VALUE-SENSITIVE DESIGN AND NANOTECHNOLOGY

1. National Center of Health Statistics, Center for Disease Control and Prevention, "Life Expectancy at Birth, at 65 Years of Age, at 75 Years of Age, according to Race and Sex: United States," 2003, http://www.cdc.gov/nchs/data/hus/tables/2003/03hus027.pdf (accessed June 7, 2010); National Center for Health Statistics, Center for Disease Control and Prevention, "Deaths: Preliminary Data for 2005," 2007, http://www.cdc.gov/nchs/prodcuts/pubs/ubd/hestats/prelimdeaths05/prelimdeaths05.htm (accessed June 7, 2010).

2. P. Richerson and R. Boyd, *Not by Genes Alone: How Culture Transformed Human Evolution* (Chicago: University of Chicago Press, 2005).

3. H. Jonas. *The Imperative of Responsibility* (Chicago: University of Chicago Press, 2003).

4. National Cancer Institute, Alliance for Nanotechnology in Cancer, "Programs," http://www.nano.cancer.gov/action/programs/ccne_subjects.asp (accessed June 7, 2010).

5. A similar approach has been used for a low-cost, high-reliability screen for blood type. M. S. Khan et al., "Paper Diagnostic for Instantaneous Blood Typing," *Analytical Chemistry* 82, no.10 (2010): 4158–64.

6. National Cancer Institute, Alliance for Nanotechnology in Cancer, "Programs."

7. National Center for Health Statistics (2007, 2003).

8. Median household income in the United States in 2008 was $52,029. US Census Bureau, "USA Quickfacts," http://www.quickfacts.census.gov/qfd/states/00000.html (accessed June 7, 2010).

9. United Nations, "Millennium Development Goals Report 2007," http://www.un.org/millenniumgoals/pdf/mdg2007.pdf (accessed June 7, 2010); United Nations, "Millenium Development Goals Report 2006,"

http://www.mdgs.un.org/unsd/mdg/Resources/Static/Products/Progress 2006/MDGReport2006.pdf (accessed June 7, 2010).

10. National Cancer Institute, President's Cancer Panel, "Reducing Environmental Cancer: What We Can Do Now," 2010, http://www.deainfo .nci.nih.gov/advisory/pcp/pcp08-09rpt/PCP_Report_08-09_508.pdf (accessed June 7, 2010).

11. D. Collingridge, *The Social Control of Technology* (New York: St. Martin's, 1980).

12. One Laptop per Child, "Mission Statement," http://www.laptop .org/en/vision/index.shtml (accessed June 7, 2010)

13. D. Talbot, "One Tablet Per Child," Technology Review, http://www .technologyreview.com/computing/25482/ (accessed June 7, 2010).

14. A criticism of the initiative is that it aims to use an expensive, complex, unproven technology to address a problem that can more efficiently and reliably be addressed using proven approaches, e.g., building schools, stocking libraries, hiring teachers, and making school free. Others have suggested that a focus on nutrition and clean water would do more to increase learning and creating (since it would reduce illness and promote development) than would this project.

15. M. J. van den Hoven, "Nanotechnology and Privacy: The Instructive Case of RFID," *International Journal of Applied Philosophy* 20, no. 2 (2006): 215–28.

16. R. Rodrigues, "The Implications of High-Rate Nanomanufacturing on Society and Personal Privacy," *Bulletin of Science, Technology, and Society* 26 no. 1 (2006): 38–45.

17. D. Guston and D. Sarewitz, "Real-Time Technology Assessment," *Technology in Society* 24, no. 1 (2001): 93–109.

CHAPTER 14: DEBATING NANOETHICS: US PUBLIC PERCEPTIONS OF NANOTECH APPLICATIONS FOR ENERGY AND ENVIRONMENT

1. This work is supported by the National Science Foundation (NSF) through grant #0824042 to the first author and Coop. Agreement No. SES 0531184 to the Center for Nanotechnology in Society at UCSB. Any opinions, findings, and conclusions or recommendations expressed in this material are those of the authors and do not necessarily reflect the views of the

NSF. The researchers would like to acknowledge and thank the following people for their contributions to the research on which this paper is based: 2007 workshops—Karl Bryant (SUNY New Paltz), Nick Pidgeon (Cardiff University, UK), Tee Rogers-Hayden (University of East Anglia, UK), Hillary Haldane (Quinnipiac University), Joseph Summers (MIT), and Terre Satterfield and Milind Kandlikar (University of British Columbia); 2009 workshops—Indy Hurt (UCSB), and Julie Whirlow (UCSB).

2. Term drawn from NSF Nano 2 "Revisioning" conference for the National Nanotechnology Initiative (NNI), convened by Mihail Roco in Evanston, Illinois, March 9–10, 2010. This gloss for ethical, legal, and societal issues (ELSI) reflects the distillation of non-environmental, health, and safety concerns. Initial comments on which this chapter is based were contributed in Session 13 by the first author.

3. The other main example of this in the US is the ELSI component of the Human Genome Project, whose funding as a proportion of overall R&D funding (3–5 percent) exceeded that of Nano ELSI (see Human Genome Project, "Ethical Legal and Social Issues," http://www.ornl.gov/sci/techresources/Human_Genome/elsi/elsi.shtml [accessed June 27, 2010]).

4. The work presented in this paper is one component of the research of the NSF Center for Nanotechnology in Society at University of California at Santa Barbara, a national center funded exclusively to conduct research, education, and outreach on the societal dimensions of nanotechnologies.

5. Cyrus Mody and Patrick W. McCray, "Big Whig History and Nano Narratives: Effective Innovation Policy Needs the Historical Dimension," *Science Progress*, 2009, http://www.scienceprogress.org/2009/04/big-whig-history-and-nano-narratives (accessed June 27, 2010).

6. David Kreibel, "Hazard Assessment: Presentation in Hazard Assessment Panel, Epidemiological Research," National Institute for Occupational Safety and Health (NIOSH) Nanomaterials and Worker Health Conference, Keystone, CO, 2010.

7. N. Choi, G. Ramachandran, and M. Kindlikar, "The Impact of Toxicity Testing Costs on Nanomaterial Regulation," *Environmental Science and Technology* 49, no. 9 (2009): 3030–34; Alexis D. Ostrowski et al., "Nanotechnology: Characterizing the Scientific Literature 2000–2007," *Journal of Nanoparticle Research* 11 (2009): 251–57.

8. Richard P. Appelbaum and Richard A. Parker, "China's Bid to Become a Global Nanotech Leader: Advancing Nanotechnology through State-Led Programs and International Collaborations," *Science and Public*

Policy 35, no. 5 (2008): 319–34; CNS-UCSB, "Nanoequity: International Conference Held at the Woodrow Wilson Center for International Scholars, Washington, DC," http://www.nanoequity2009.cns.ucsb.edu/ (accessed June 27, 2010).

9. Terre A. Satterfield, C. K. Mertz, and Paul Slovic, "Discrimination, Vulnerability, and Justice in the Face of Risk," *Risk Analysis* 24, no. 1 (2004): 115–29.

10. Cf. Nick Pidgeon and Tee Rogers-Hayden, "Opening up Nanotechnology Dialogue with the Publics: Moving Beyond Risk Debates to 'Upstream Engagement,'" *Health, Risk & Society* 9, no. 2 (2007): 191–210; Tee Rogers-Hayden, Nick Pidgeon, and A. Mohr, "Engaging with Nanotechnologies—Engaging Differently," *NanoEthics*, 1, no. 2 (2007): 143–54.

11. National Research Council, "A Matter of Size: Triennial Review of the National Nanotechnology Initiative" (Washington, DC: National Academies Press, 2006), http://www.books.nap.edu/catalog.php?record_id=11752 (accessed June 27, 2010).

12. "The Idea of Progress: Onwards and Upwards," *Economist*, December 19, 2009.

13. Susan Nieman, *Moral Clarity: A Guide for Grown-up Idealists* (New York: Harcourt, 2008).

14. Mihail C. Roco, "Forward," in *Deliberative Democracy and Nanotechnology*, ed. Fritz Allhoff, et al. (New Jersey: John Wiley & Sons Inc, 2007): p. xii.

15. John Gastil, *Political Communication and Deliberation* (Los Angeles: Sage, 2008).

16. Ibid; Patricia Moy and John Gastil, "Predicting Deliberative Conversation: The Impact of Discussion Networks, Media Use, and Political Cognitions," *Political Communication* 23, no. 4 (2006): 443–60.

17. Colin Farrelly, "Deliberative Democracy and Nanotechnology," in *Nanoethics: The Ethical and Social Implications of Nanotechnology*, ed. Fritz Allhoff, Patrick Lin, James Moor, and John Weckert (New Jersey: John Wiley & Sons, 2007), p. 216 (author's emphasis).

18. R. Gregory, "Using Stakeholder Values to Make Smarter Environmental Management Decisions," *Environment* 42, no. 5 (2000): 34–44; R. Gregory, "Incorporating Value Tradeoffs into Community-Based Environmental Risk Decisions," *Environmental Values* 11 (2002): 461–88.

19. Stephen Macedo, *Deliberative Politics: Essays on Democracy and Disagreement* (New York: Oxford University Press, 1999); Katherine C. Walsh, *Talking about Politics: Informal Groups and Social Identity in American Life* (Chicago: University of Chicago Press, 2004); Sue Wilkinson, "Focus

Groups: A Feminist Method," *Psychology of Women Quarterly* 23, no. 2 (1999): 221–44.

20. Cf. Bruce V. Lewenstein, "What Counts as a 'Social and Ethical Issue' in Nanotechnology?" in *Nanotechnology Challenges: Implications for Philosophy, Ethics, and Society* (Singapore: World Scientific, 2006).

21. Paul Slovic, "Perceived Risk, Trust, and Democracy," *Risk Analysis* 16, no. 6 (1993): 675–82.

22. William R. Freudenburg, "Risk and Recreancy: Weber, the Division of Labor, and the Rationality of Risk Perceptions," *Social Forces* 71, no. 4 (1993): 909–932.

23. Nick Pidgeon et al., "Deliberating the Risks of Nanotechnologies for Energy and Health Applications in the United States and United Kingdom," *Nature Nanotechnology* 4, no. 2 (2009): 95–98.

24. See Fritz Allhoff, Patrick Lin, and Daniel Moor, *What Is Nanotechnology and Why Does It Matter?* (Malden, MA: Wiley-Blackwell, 2010). See chapter 7 for a discussion of equity and access issues in relation to water purification, solar energy, and medical nanotech applications.

25. Robert D. Bullard and Glenn S. Johnson, "Environmental Justice: Grassroots Activism and Its Impact on Public Policy Decision Making," *Journal of Social Issues* 56, no. 3 (2000): 555–78; Rae Zimmerman, "Social Equity and Environmental Risk," *Risk Analysis* 13, no. 6 (1993): 649–66; Mody and McCray, "Big Whig History and Nano Narratives."

26. Cf. Jennifer Kuzma and John C. Besley, "Ethics of Risk Analysis and Regulatory Review: From Bio- to Nanotechnology," *NanoEthics* 2 (2008): 149–62; Katherine McComas, John C. Besley, and Zheng Yang, "Risky Business: Perceived Behavior of Local Scientists and Community Support for Their Research," *Risk Analysis* 28, no. 6 (2008): 1539–52.

27. Satterfield, Mertz, and Slovic, "Discrimination, Vulnerability, and Justice."

28. Cf. Paul Slovic, *The Perception of Risk* (London: Routledge, 2000).

29. Satterfield, Mertz, and Slovic, "Discrimination, Vulnerability, and Justice."

30. National Institute for Occupational Safety and Health (NIOSH), "Progress toward Safe Nanotechnology in the Workplace: A Report from the Niosh Nanotechnology Research Center," NIOSH Publication No. 2010-104, November 2009, http://www.198.246.98.21/niosh/docs/2010-104/ (accessed June 27, 2010).

31. Pidgeon and Rogers-Hayden, "Opening up Nanotechnology Dialogue with the Publics."

32. J. Wilsdon and R. Willis, *See-Through Science: Why Public Engagement Needs to Improve Upstream* (London: Demos, 2004).

33. For purposes of ethical research conduct and effective group interaction, including persons with less than a high school degree in activities that require ability to read texts and verbally discuss and debate ideas in English may be uncomfortable for them and counterproductive for the research (Shawn W. Rosenberg, "An Introduction: Theoretical Perspectives and Empirical Research on Deliberative Democracy," in *Deliberation, Participation, and Democracy: Can the People Govern?* ed. Shawn W. Rosenberg [New York: Palgrave MacMillan, 2007]).

34. Peter Bennett et al., *Risk Communication and Public Health* (New York: Oxford University Press, 2010).

35. Cf. Mary Douglas, *Risk and Blame: Essays in Cultural Theory* (London: Routledge, 1992); Barbara Herr Harthorn and Laury Oaks, *Risk, Culture, and Health Inequality: Shifting Perceptions of Danger and Blame* (Westport, CT: Praeger/Greenwood, 2003).

36. Cf. Slovic, "Perceived Risk, Trust, and Democracy."

37. Thomas Dietz and Paul Stern, ed., *Public Participation in Environmental Assessment and Decision Making* (Washington, DC: National Academies Press, 2008); Paul Stern and Harvey Fineberg, *Understanding Risk: Informing Decisions in a Democratic Society* (Washington, DC: National Academies Press, 1996).

38. All names in examples are pseudonyms; all demographics are accurate.

39. Pidgeon et al., "Deliberating the Risks of Nantechnologies for Energy and Health Applications."

40. Sarah Davies, Phil Macnaghten, and Matthew Kearnes, *Reconfiguring Responsibility: Deepening Debate on Nanotechnology* (Durham, UK: Durham University, 2009), http://www.geography.dur.ac.uk/projects/deepen/NewsandEvents/tabid/2903/Default.aspx (accessed June 27, 2010).

41. Gastil, *Political Communication and Deliberation*; Moy and Gastil, "Predicting Deliberative Conversation."

42. Pidgeon et al., "Deliberating the Risks of Nantechnologies for Energy and Health Applications."

43. National Institute for Occupational Safety and Health (NIOSH), "Current Intelligence Bulletin 60: Interim Guidance for Medical Screening and Hazard Surveillance for Workers Potentially Exposed to Engineered Nanoparticles," 2009, http://www.cdc.gov/niosh/docs/2009-116/ (accessed June 27, 2010).

44. Paul Slovic, "Perception of Risk," *Science* 236 (1987): 280–85; Slovic, "Perceived Risk, Trust, and Democracy."

45. R. Gregory and T. Satterfield, "Beyond Perception: The Experience of Risk and Stigma in Community Contexts"; Satterfield, Mertz, and Slovic, "Discrimination, Vulnerability, and Justice."

46. Paul Slovic, "Perceived Risk, Trust, and Democracy."